Genus *Rheum* (Polygonaceae)

Genus *Rheum* (Polygonaceae)
A Global Perspective

Shahzad A. Pandith
Mohd Ishfaq Khan

CRC Press
Taylor & Francis Group
Boca Raton London New York

CRC Press is an imprint of the
Taylor & Francis Group, an **informa** business

First edition published 2022
by CRC Press
6000 Broken Sound Parkway NW, Suite 300, Boca Raton, FL 33487-2742

and by CRC Press
2 Park Square, Milton Park, Abingdon, Oxon, OX14 4RN

© 2022 Taylor & Francis Group, LLC
CRC Press is an imprint of Taylor & Francis Group, LLC

Reasonable efforts have been made to publish reliable data and information, but the author and publisher cannot assume responsibility for the validity of all materials or the consequences of their use. The authors and publishers have attempted to trace the copyright holders of all material reproduced in this publication and apologize to copyright holders if permission to publish in this form has not been obtained. If any copyright material has not been acknowledged please write and let us know so we may rectify in any future reprint.

Except as permitted under U.S. Copyright Law, no part of this book may be reprinted, reproduced, transmitted, or utilized in any form by any electronic, mechanical, or other means, now known or hereafter invented, including photocopying, microfilming, and recording, or in any information storage or retrieval system, without written permission from the publishers.

For permission to photocopy or use material electronically from this work, access www.copyright.com or contact the Copyright Clearance Center, Inc. (CCC), 222 Rosewood Drive, Danvers, MA 01923, 978-750-8400. For works that are not available on CCC please contact mpkbookspermissions@tandf.co.uk

Trademark notice: Product or corporate names may be trademarks or registered trademarks and are used only for identification and explanation without intent to infringe.

ISBN: 9780367355760 (hbk)
ISBN: 9780429340390 (ebk)
ISBN: 9781032058429 (pbk)

DOI: 10.1201/9780429340390

Typeset in Times
by Deanta Global Publishing Services, Chennai, India

Contents

Preface vii
Authors ix

Chapter 1 Introduction 1

          1.1   Introduction 1
          References 4

Chapter 2 Botany and Ecology 7

          2.1   Occurrence and Distribution 7
          2.2   Taxonomy and Classification 7
          2.3   Morphology 14
          2.4   Synonyms and Vernacular Names 25
          2.5   Pollen Studies 25
          2.6   Ecology 33
          References 34

Chapter 3 Traditional Uses 39

          3.1   Rhubarb as a Food Plant 42
          References 44

Chapter 4 Phytochemistry 47

          4.1   Phenolics 48
          4.2   Anthraquinones 61
          4.3   Stilbenoids 62
          4.4   Other Constituents 65
                  4.4.1   Anthraglycosides 70
                  4.4.2   Dianthrones 71
          References 71

Chapter 5 Pharmacology 75

          5.1   Antimicrobial Activity 75
          5.2   Anticancer Activity 77
          5.3   Antidiabetic Activity 80
          5.4   Anti-Inflammatory Activity 83
          5.5   Antioxidative Activity 84
          5.6   Immunoenhancing Activity 87
          5.7   Nephroprotective Activity 87

vi   Contents

	5.8	Hepatoprotective Activity 90
	5.9	Miscellaneous Activities 92
	References 94

Chapter 6   Molecular Aspects 107

	6.1	Cytogenetics 107
	6.2	Genetics 110
		6.2.1 Rhubarb in the Wild 111
		6.2.2 Rhubarb Cultivars 114
	6.3	Biotechnological Interventions 119
	References 126

Chapter 7   Conservation 135

	7.1	Threat Status 135
	7.2	*In vitro* Propagation Studies as a Conservation Measure 136
	References 138

Chapter 8   Conclusions and Future Prospects 141

	References 146

Preface

Plants, sessile but promising organisms, have been well-known for benefiting human existence in the form of a variety of natural products for ages. Among this vast reservoir of natural economic wealth, rhubarb (Polygonaceae) is one of the important medicinal herbs with an intricate historical background and an immense value for the associated societies for centuries. Rhubarb, a significant object of botanical, commercial, and horticultural interest, is presently represented by about 60 extant species occurring across Asian and European countries with a major distribution found in the mountainous regions of the Qinghai-Tibetan Plateau in China. This perennial species is documented for use in various traditional medical systems from varied cultures across different regions as one of the most sought-after crude drugs due to its mildness, efficacy, and lack of unwanted side effects. Indeed, in therapeutic history, there has probably been no medicine which has provided better relief to more people than the rhizome/root powder made from medicinal rhubarb (*Rheum officinale*). Moreover, there has been remarkable interest from researchers around the world in some of the highly medicinal herbs from this genus which, although not limited to, include *R. australe*, *R. tanguticum*, *R. palmatum*, and *R. officinale*, among others. Pertinently, the remedying properties of *Rheum* are attributed to a group of diverse biologically active secondary chemical constituents, predominantly anthraquinones, and stilbenoids to a lesser extent, as well as the dietary flavonoids known for their putative health benefits. But unfortunately, human greed has prevailed and rendered some of the important species from this Polygonaceous herb threatened in the natural stands, causing the need for their conservation. Further, with the advancement of the state-of-the-art scientific techniques, current and related research is mainly focused on various promising metabolic pathways from medicinal plants like *Rheum* to decipher the biosynthesis of their key bioactive chemical constituents, which may help in modulating the specific pathways in alternate (microbial) hosts while preserving the natural wild stock.

This book is literally the biography of the rhubarb plant and presents a global perspective of this important medicinal herb while covering almost every aspect, ranging from its history to the application of contemporary biotechnological interventions. It provides updated information on nearly all aspects of the plant covering varied forms of related literature since the 18th century, and include:

- The historical background detailing its earlier trade and commerce;
- Global distribution giving an idea of the source of its origin and dissemination;
- Botany for classical taxonomic aspects;
- Ecology to understand the ecological and evolutionary implications;
- Ethnobotany, phytochemistry and pharmacology vis-à-vis therapeutic potential;

- Genetic diversity with regard to the polymorphism and ploidy status gained across varied geographies;
- Biotechnological invasions to assess the success level we have achieved so far in accepting and utilizing rhubarb as a suitable drug reservoir; and
- *In vitro* propagation studies as a conservation and exploitation (at industrial scale) measure.

We are hopeful that it will prove to be of great help to the researchers working on rhubarb.

Shahzad A. Pandith
Mohd Ishfaq Khan

Authors

Shahzad A. Pandith received his MSc in botany from the University of Kashmir, Srinagar, India, in 2009, and PhD in plant molecular biology from CSIR Indian Institute of Integrative Medicine, Jammu, India, in 2017. He has since served in the Department of Botany, University of Kashmir as DST-INSPIRE Faculty (position equivalent to IITian assistant professor). In this capacity, he supervises a DST-sponsored project on one of the important medicinal herbs from the northwestern Himalayas, *Rheum*. In general, Pandith seeks to use various "omics" approaches to elucidate and understand the basic biology of plants, important pathways operating in them and their modulation, rate-limiting pathway genes, and the regulatory factors controlling their constitutive/inducible expression vis-à-vis ecological/environmental factors and/or stress conditions. He has published about 15 research papers in highly reputed peer-reviewed international journals with an average impact of around three per paper. In addition, he is an active reviewer of many reputed journals from well-known publishing houses like Springer, Elsevier, and Brill, etc. Shahzad has many national-level exams and awards to his credit, which include the Joint CSIR-UGC JRF and ICAR-ASRB NET exams, and the highly prestigious DST-INSA INSPIRE Faculty award.

Mohd Ishfaq Khan holds a BSc degree from the University of Kashmir, Srinagar, India, and received his MSc degree in botany from the same institution in 2015. Since 2017, he is actively engaged in a DST-sponsored project at the University of Kashmir. His active areas of research include DNA barcoding, genetics, taxonomy, and eco-physiology which he uses to understand and resolve the genetic/taxonomic complexity in high-value medicinal herbs from the northwestern Himalayas. He has some good research articles to his credit. In addition, he has qualified in many competitive exams at state and national level including the Graduate Aptitude Test in Engineering (GATE)-2019 in the subject of life sciences. Khan also holds a bachelor's degree in education and has been a good teacher-cum-motivator for students at higher secondary level.

1 Introduction

1.1 INTRODUCTION

Throughout the ages, nature has been the source for basic human needs. Plants, in particular, have formed the basis of human existence from time immemorial, probably (as medicine) from the middle Paleolithic age some 60,000 years ago (Solecki 1975). Since then, humans have been using plants as food for themselves and for their livestock, as well as for manufacturing various products and using them for different purposes. Owing to this close historical association between plants and humankind, early man, over a period of time, started distinguishing plants with medical efficacy from the ones with varied uses and which play a significant role in maintaining ecological balance. This ethnomedicinal knowledge, transmitted down the generations through oral traditions, proximately resulted in the generation and documentation of this information in the form of various widely accepted and practiced traditional medical systems, viz. Unani, Ayurveda, Siddha, naturopathic, Tibetan, homeopathic, and Chinese systems, among others. Ultimately, with centuries/decades of hard investment, experience, and the advances in research made in the field, a diverse range of novel drugs with plant origin to cure a wide array of ailments across borders came into existence and is also growing at an accelerating pace in light of modern scientific technology. In fact, as per the estimation of the World Health Organization, about 75% of the population of Asian and African countries still rely on conventional and traditional methods for the treatment of various diseases (Wani et al. 2013; Pandith et al. 2014). These preventative measures mostly involve the use of herbal extracts which are cheap and offer comparatively high safety and fewer side effects. In the last few decades, plant-derived natural products have found use as efficient chemotherapeutic and/or chemopreventive agents for the effective treatment of a vast range of diseases. Pertinently, phytomedicine offers a great deal to the health system, as presently about 25% of pharmaceutical prescriptions in the United States contain at least one plant-derived ingredient (Pandith et al. 2014). Basically, the metabolomes (chemical profile) of medicinal plants serve as a valuable natural resource for pharmaceuticals and provide room for renewed attention from both practical and scientific viewpoints for the evidence-based development of new phytotherapeutics and nutraceuticals. Moreover, new screening approaches are being developed to improve the ease with which natural products (NPs) can be used in drug discovery and development. It is also anticipated that more efficient and effective application of NPs will improve the drug discovery process.

As per the literature survey, one recent investigation suggests the number of angiosperm species to be around 450,000, although 10–20% of this number is

DOI: 10.1201/9780429340390-1

believed to be unknown to the scientific world. About two-thirds of these species occur within the tropics with a third of them (including the undescribed ones) facing the risk of extinction (Corlett 2016; Pimm and Joppa 2015). Among this vast reservoir of angiosperm taxa, rhubarb (Polygonaceae) is one of the significant plant species with an intricate historical background and an immense value to specific societies over centuries. In therapeutic history, there has probably been no medicine which has realized better relief to more people than the rhizome/root powder made from medicinal rhubarb (*R. officinale*). In fact, from ancient times to the recent past, rhubarb has remained one of the most sought-after crude drugs due to its mildness, efficacy, and lack of unwanted side effects. Moreover, there was a sort of fascination with rhubarb as a much-used cathartic therapy and tonic in 18th- and 19th-century America and Europe, and a trend for its use as a foodstuff especially in 19th-century Britain and America. Indeed, numerous editions of the *Encyclopedia Britannica* delivered gave an informal indication of the rise and fall in acceptance of this amazing aperient drug (and sour vegetable/fruit) (Clifford 1992). Perceiving its immense worth in a world in need of relief from recurrent constipation events, rhubarb has attracted the attention of botanists, explorers, horticulturists, merchants, pharmacists, and physicians alike, and over centuries. Even though it is a significant object of botanical, commercial, and horticultural interest, rhubarb has been, in its eventual definition, a therapeutic medicine and has largely grabbed the attention of both clinical and theoretical physicians by actively serving principal medical needs—maintaining internal body passages open and clean while eliminating detrimental humors—in the 18th century. Further, throughout rhubarb's history, existence of the "true rhubarb"—one official species/plant which gives the finest drug—was presumed by many associated commentators. This assumption was supplemented during the 18th century when it became clear that there existed many species/varieties of this plant sharing some characteristics which make them resemble one another (Foust 2014). The Asian lands served as intermediaries for rhubarb to be imported to Europe until the Renaissance which led Europeans to reach nearer to its source, though Arabs had understood, yet not precisely, that rhubarb roots exported to Europe originated somewhere in the Chinese Empire. Authoritatively, it was erudite only in the next half of the 19th century that the rhubarb roots of high medicinal value were native to the west China highlands, northern Tibet, and southern Mongolia, whereas the species with relatively less medical efficacy occurred in Bhutan, northern India, and south Himalayan Nepal, and the ones with meager medical benefits were from Siberia, south-west Asia and south-east Europe (Clifford 1992; Foust 2014).

The genus *Rheum* (rhubarb) includes well-known traditional medicinal plants which mainly occur in mountainous regions of the Qinghai-Tibetan Plateau (QTP) and its adjacent areas (Wan et al. 2011). Owing to the diverse morphology and substantial endemism, these regions are supposed to form both the origin and diversification centers for the genus consisting of nearly 60 extant species which are perennial herbs distributed along the temperate, alpine, and sub-alpine zones on rocky, porous, and humus-rich soils at an elevation of 500–5400 m asl

Introduction

(Wan et al. 2011; Wu et al. 2003). The species of genus *Rheum* occurs across Asian (Afghanistan, Armenia, Azerbaijan, Bhutan, China, India, Iran, Iraq, Kazakhstan, Lebanon, Mongolia, Myanmar, Nepal, Pakistan, Syria, Turkey) and European (Russia) countries with 19 of them endemic to China (Wu et al. 2003). Earlier researchers mainly used the visible robust diversity to establish different infra-generic sections within this genus (Losina-Losinskaya 1936a, b; Kao and Cheng 1975; Li 1998) which started with nine sections by Losina-Losinskaya (1936a) and ended with, to date, eight sections based on six different types of pollen grains (Li 1998). Different species of rhubarb found extensive use in various traditional systems of medicine, viz. Ayurveda, Unani, Siddha, and Chinese systems, etc.

The ethnomedical uses of *Rheum* have been documented in nearly every country where it occurs. Indeed, there has been remarkable interest from researchers around the globe in some of the highly medicinal herbs from this genus which include, but are not limited to, *R. australe*, *R. tanguticum*, *R. palmatum*, and *R. officinale*, etc. These medicinally potent species of the genus *Rheum* are reportedly known for various biological properties including nephroprotective, hepatoprotective, immuno-enhancing, antioxidant, anti-inflammatory, antifungal, antidiabetic, anticancer, and antibacterial. They are also known to be used for the treatment of tumors, pain, inflammation, bleeding, gastric ulcers, tinea, dysmenorrhea, constipation, Parkinson's disease, and severe acute respiratory syndrome (SARS) (Agarwal et al. 2001; Zheng et al. 2013; Wang et al. 2012; Pandith et al. 2018, 2014; Rokaya et al. 2012). The remedying properties of *Rheum* are attributed to a group of diverse biologically active secondary chemical constituents, predominantly anthraquinones (emodin, chrysophanol, physcion, aloe-emodin, and rhein) and stilbenoids (piceatannol, resveratrol) to a lesser extent, comparatively, besides the dietary phytoconstituents (flavonoids) known for their putative health benefits. In fact, some of the recent investigations in the area tend to determine, validate, and establish the significant pharmacological efficiency of selected secondary metabolites and/or their derivatives as lead molecules for the effective treatment of various human ailments (Pandith et al. 2018, 2014).

Over the past few decades, some of the species from the genus *Rheum* have seen dwindling populations in their natural habitats due to different kinds of prevailing natural and anthropogenic pressures. More importantly, overexploitation of *Rheum* species for herbal drug preparations from natural habitats has led to a significant reduction in its populations in its natural state. Consequently, the alarming statistics of the availability of many species of high therapeutic value in nature—*R. australe*, *R. tanguticum*, *R. webbianum*, and *R. palmatum*—has led them to figure prominently among endangered plant species (Pandith et al. 2018, 2016; Rokaya et al. 2012; Chen et al. 2009; Wang et al. 2012; Rashid et al. 2014). Indeed, these alarming statistics of the availability of *Rheum* in nature call for urgent attention to arrest and reverse this trend by making strategic efforts in conservation and sustainable utilization vis-à-vis the regional economic benefits of this important medicinal herb.

4 Genus *Rheum*: A Global Perspective

This compilation on the genus *Rheum* (rhubarb) is to a greater extent the biography of the plant at global scale. The write-up covers almost every aspect of the plant from classical taxonomy to contemporary biotechnological interventions.

- This book provides up-to-date information available on its botany since the 18th century for easy location and identification of the plant;
- Ecology to understand its ecological and evolutionary implications;
- Origin and historical perspective detailing its trade and commerce;
- Global distribution giving an idea of its origin and dissemination;
- Therapeutic potential in relation to traditional uses and pharmacological efficacy;
- Phytochemistry, discussing major bioactive chemical constituents;
- Genetic and cytogenetic studies leading the way to explore the levels of genetic diversity it harbors, and to realize the ploidy status across varied geographies;
- Biotechnological intrusions to see and judge the level of success we have achieved so far in accepting and utilizing the target species as a suitable platform for potent and efficacious lead molecules for future drugs; and
- Finally, *in vitro* propagation studies holding vital significance in preserving and conserving the natural germplasm of the plant and for its industrial exploitation.

It also provides a detailed perspective for future studies to conserve and make sustainable use of this economic and medicinal wealth at an industrial and commercial scale.

REFERENCES

Agarwal SK, Singh SS, Lakshmi V, Verma S, Kumar S (2001) *Chemistry and Pharmacology of Rhubarb (Rheum species)—A Review*. New Delhi, India: NISCAIR-CSIR.

Chen F, Wang A, Chen K, Wan D, Liu J (2009) Genetic diversity and population structure of the endangered and medically important Rheum tanguticum (Polygonaceae) revealed by SSR markers. *Biochemical Systematics and Ecology* 37(5):613–621.

Clifford M (1992) *Rhubarb: The Wondrous Drug*. Princeton, NJ: Princeton University Press.

Corlett RT (2016) Plant diversity in a changing world: Status, trends, and conservation needs. *Plant Diversity* 38(1):10–16.

Foust CM (2014) *Rhubarb: The Wondrous Drug*, Volume 191. Princeton, NJ: Princeton University Press.

Kao T, Cheng C-Y (1975) Synopsis of the Chinese rheum. *Acta Phytotaxonom sinica* 31:1–25.

Li A (1998) *Flora Reipublicae Popularis Sinicae: Tomus 25 (1). Angiospermae, Dicotyledoneae, Polygonaceae*. Beijing: Science Press:237.-illus.. ISBN 703006450X Ch Icones, Keys. Geog.

Losina-Losinskaya A (1936a) The Genus Rheum and Its Species. In: *Acta Instituti Botanici Academiae Scientiarum URSS, Unionis Rerum, Publicarum Soveticarum Series, Socialisticarum Series 1, Fasciculus 3:67-141*: Moscow, Russia.

Introduction

Lozina-Lozinskaja A (1936b) Sistematiceskij obzor dikorastuscich vidov roda Rheum L. In: *Trudy Botanicheskogo Instituta Akademii Nauk SSSR. Series 1* Moscow, Russia: 67–141.

Pandith SA, Dar RA, Lattoo SK, Shah MA, Reshi ZA (2018) Rheum australe, an endangered high-value medicinal herb of North Western Himalayas: A review of its botany, ethnomedical uses, phytochemistry and pharmacology. *Phytochemistry Reviews : Proceedings of the Phytochemical Society of Europe* 17(3):1–37.

Pandith SA, Dhar N, Rana S, Bhat WW, Kushwaha M, Gupta AP, Shah MA, Vishwakarma R, Lattoo SK (2016) Characterization and functional promiscuity of two divergent paralogs of Type III plant polyketide synthases from Rheum emodi Wall ex. Meissn. *Plant Physiology* 171(4):2599–2619.

Pandith SA, Hussain A, Bhat WW, Dhar N, Qazi AK, Rana S, Razdan S, Wani TA, Shah MA, Bedi Y, Hamid A, Lattoo SK (2014) Evaluation of anthraquinones from Himalayan rhubarb (Rheum emodi Wall. ex Meissn.) as antiproliferative agents. *South African Journal of Botany* 95:1–8.

Pimm SL, Joppa LN (2015) How many plant species are there, where are they, and at what rate are they going extinct? *Annals of the Missouri Botanical Garden* 100(3):170–176.

Rashid S, Kaloo ZA, Singh S, Bashir I (2014) Callus induction and shoot regeneration from rhizome explants of Rheum webbianum Royle-a threatened medicinal plant growing in Kashmir Himalaya. *Journal of Scientific and Innovative Research* 3(5):515–518.

Rokaya MB, Münzbergová Z, Timsina B, Bhattarai KR (2012) Rheum australe D. Don: A review of its botany, ethnobotany, phytochemistry and pharmacology. *Journal of Ethnopharmacology* 141(3):761–774.

Solecki RS (1975) Shanidar IV, a Neanderthal flower burial in northern Iraq. *Science* 190(4217):880–881.

Wan D, Wang A, Zhang X, Wang Z, Li Z (2011) Gene duplication and adaptive evolution of the CHS-like genes within the genus Rheum (Polygonaceae). *Biochemical Systematics and Ecology* 39(4–6):651–659.

Wang X, Yang R, Feng S, Hou X, Zhang Y, Li Y, Ren Y (2012) Genetic variation in Rheum palmatum and Rheum tanguticum (Polygonaceae), two medicinally and endemic species in China using ISSR markers. *PLOS ONE* 7(12):e51667.

Wani BA, Ramamoorthy D, Rather MA, Arumugam N, Qazi AK, Majeed R, Hamid A, Ganie SA, Ganai BA, Anand R, Gupta AP (2013) Induction of apoptosis in human pancreatic MiaPaCa-2 cells through the loss of mitochondrial membrane potential ($\Delta\Psi$m) by Gentiana kurroo root extract and LC-ESI-MS analysis of its principal constituents. *Phytomedicine* 20(8–9):723–733.

Wu Z, Raven PH, Hong D (2003) Flora of China. Volume 5 *Ulmaceae through Basellaceae*. Beijing, China: Science Press.

Zheng Q-X, Wu H-F, Jian G, Nan H-J, Chen S-L, Yang J-S, Xu X-D (2013) Review of rhubarbs: Chemistry and pharmacology. *Chinese Herbal Medicines* 5(1):9–32.

2 Botany and Ecology

2.1 OCCURRENCE AND DISTRIBUTION

The genus *Rheum* consisting of about 60 extant species is typically limited to Asia (Airy Shaw 1973; Gohil and Rather 1986) with central and northern parts of Asia as the center of its distribution (Lozina-Lozinskaja 1936a, b), although literature shows a few records of it occurring outside these regions as well (Libert and Englund 1989). The species are distributed along the temperate, alpine, and sub-alpine zones on rocky, porous, and humus-rich soils at an elevation of 500 m (*R. tataricum*) to 5400 m (*R. rhomboideum*) (Wu et al. 2003). As per the available fossil records, the occurrence of Polygonaceae dates back to the Paleocene epoch (Muller 1981). Further, based on the molecular clock hypothesis, the divergence of *Rheum* and its sister groups dates to the Miocene epoch (Sun et al. 2012; Wang et al. 2012). In the Indian subcontinent, ten species of the genus are known to exist (Hooker 1897; Santapau and Henry 1973) with six (*R. moorcroftianum* mentioned as a synonym for *R. spiciforme* Royle) of them reported from Kashmir—covering the entire erstwhile state of Jammu and Kashmir (excluding the lower Shivaliks in Jammu)—which is currently part of India, Pakistan, and China (Stewart et al. 1972). However, a recent report (Srivastava 2014) states that only eight species of the genus exist in India and include *Rheum nobile* Hook. f. & Thomson, *R. globulosum* Gage, *R. spiciforme* Royle, *R. moorcroftianum* Royle, *R. tibeticum* Maxim. ex Hook. f., *R. acuminatum* Hook. f. & Thomson, *R. australe* D. Don, and *R. webbianum* Royle. The species of *Rheum* occur across Asian and European countries with 19 of them endemic to China (Wu et al. 2003) with other countries having a smaller share; for instance, only three *Rheum* species (*R. undulatum*, *R. rhaponticum*, and *R. palmatum*) are known to grow in Poland (Medynska and Smolarz 2005). The occurrence and distributional range of different species of *Rheum* are listed in detail in Table 2.1.

2.2 TAXONOMY AND CLASSIFICATION

Polygonaceae Juss., commonly known as knotweed or the smartweed-buckwheat family, is a large and diverse family of dicotyledonous angiosperms exhibiting extensive plasticity in growth forms (herbs, lianas, shrubs, vines, or trees with often swollen nodes) with unique morphological characters (quincuncial aestivation, orthotropous ovules and trigonal achenes, ocrea) (Sanchez and Kron 2009). Genus type "*Polygonum*" forms the basis for the name "Polygonaceae" and was initially used in 1789 by Antoine Laurent de Jussieu in his book *Genera Plantarum* (de Jussieu 1789). It is derived from Greek: "poly" means "many," and "goni" means "knee/joint," which refer to the several nodes on the stem/branches. The family

DOI: 10.1201/9780429340390-2

TABLE 2.1

Global Distribution Range of Different Species of the Genus *Rheum* Linnaeus

S. No.	Species	Regions of Occurrence	Altitude (m asl)	Reference
1	*R. acuminatum* Hook. f. & Thomson	China, Tibet, Bhutan, India (Kashmir, Sikkim), Myanmar, Nepal	2800–4000	FOC Vol. 5, p. 346
2	*R. nobile* Hook	Native to the Himalaya, Afghanistan, Pakistan, India, Nepal, Sikkim, Bhutan, China (Tibet), Myanmar	4000–4800	FOC Vol. 5, p. 350
3	*R. ribes* Linnaeus	Turkey, Syria, Lebanon, Iraq, Iran, Azerbaijan, Armenia, Afghanistan, Pakistan	1000–4000	FOP Vol. 205, p. 212
4	*R. tibeticum* Maxim. ex Hook	Afghanistan, Pakistan, India (Ladakh), China (Xizang)	2500–4000	FOP Vol. 205, p. 213
5	*R. moorcroftianum* Royle	From India (Kumaon) to Nepal	1000–4000	FOP Vol. 205, p. 212
6	*R. wittrockii* Lundstr.	Pakistan and Central Asia	1000–4000	FOP Vol. 205, p. 213
7	*R. australe* D. Don	Pakistan, India, Nepal	1500–4000	FOP Vol. 205, p. 212
8	*R. webbianum* Royle	China, India (NW), Nepal, Pakistan	3500–3600	FOC Vol. 5, p. 343
9	*R. globulosum* Gage, Bull. Misc	China, India (Sikkim)	4500–5000	FOC Vol. 5, p. 350
10	*R. reticulatum* Losinskaja, Trudy	China, Kazakhstan, Kyrgyzstan, Tajikistan	2900–4200	FOC Vol. 5, p. 350
11	*R. maculatum* C. Y. Cheng & T. C. Kao	China	No data found	FOC Vol. 5, p. 346
12	*R. compactum* Linnaeus	China, Kazakhstan, Mongolia, Russia (Far East, Siberia)	2000	FOC Vol. 5, p. 345
13	*R. alexandrae* Batalin, Trudy Imp	China (W Sichuan, E Xizang, NW Yunnan)	3000–4600	FOC Vol. 5, p. 350
14	*R. przewalskyi* Losinskaja, Trudy	China (Gansu, Qinghai, NW Sichuan)	1500–5000	FOC Vol. 5, p. 349
15	*R. rhizostachyum* Schrenk	China (Xinjiang), Kazakhstan	2600–4200	FOC Vol. 5, p. 350

(Continued)

Botany and Ecology

TABLE 2.1 (CONTINUED)
Global Distribution Range of Different Species of the Genus *Rheum* Linnaeus

S. No.	Species	Regions of Occurrence	Altitude (m asl)	Reference
16	*R. rhomboideum* Losinskaja	China (C and E Xizang)	4700–5400	FOC Vol. 5, p. 349
17	*R. tataricum* Linnaeus	China (W Xinjiang), Afghanistan, Kazakhstan, Russia (European part)	500–1000	FOC Vol. 5, p. 349
18	*R. uninerve* Maximowicz	China (Gansu, Nei Mongol, E Qinghai), Mongolia	1100–2300	FOC Vol. 5, p. 348
19	*R. nanum* Siev. ex Pall.	China (Gansu, C and W Nei Mongol, NE Xinjiang), Kazakhstan, Mongolia, Russia (W Siberia)	700–2000	FOC Vol. 5, p. 348
20	*R. sublanceolatum* C. Y. Cheng & T. C. Kao	China (Gansu, Qinghai, Xinjiang)	2400–3000	FOC Vol. 5, p. 347
21	*R. racemiferum* Maximowicz	China (Gansu, Nei Mongol, Ningxia), Mongolia	1300–2000	FOC Vol. 5, p. 347
22	*R. inopinatum* Prain	China (C and S Xizang)	4000–4200	FOC Vol. 5, p. 347
23	*R. subacaule* Samuelsson	China (W Sichuan)	3500–4300	FOC Vol. 5, p. 348
24	*R. pumilum* Maximowicz	China (Gansu, Qinghai, Sichuan, Xizang)	2800–4500	FOC Vol. 5, p. 347
25	*R. delavayi* Franchet	China (W Sichuan, N Yunnan), Bhutan, Nepal	3000-4800	FOC Vol. 5, p. 348
26	*R. yunnanense* Samuelsson	China (NW Yunnan), Myanmar	4000	FOC Vol. 5, p. 346
27	*R. kialense* Franchet	China (Gansu, Sichuan, Yunnan)	2800–3900	FOC Vol. 5, p. 347
28	*R. tanguticum* Maxim. ex Balf.	China (Gansu, Qinghai, Shaanxi, Xizang)	1600–3000	FOC Vol. 5, p. 345
29	*R. laciniatum* Prain	China (N Sichuan)	3000	FOC Vol. 5, p. 346
30	*R. hotaoense* C. Y. Cheng & T. C. Kao	China (Gansu, Shaanxi, Shanxi)	1000–1800	FOC Vol. 5, p. 344
31	*R. altaicum* Losinskaja	China (N Xinjiang), Kazakhstan, Mongolia, Russia	1900–2400	FOC Vol. 5, p. 344

(Continued)

TABLE 2.1 (CONTINUED)
Global Distribution Range of Different Species of the Genus *Rheum* Linnaeus

S. No.	Species	Regions of Occurrence	Altitude (m asl)	Reference
32	*R. glabricaule* Samuelsson	China (Gansu)	3000–3500	FOC Vol. 5, p. 345
33	*R. forrestii* Diels	China (Sichuan, Xizang, Yunnan)	3000	FOC Vol. 5, p. 344
34	*R. likiangense* Sam.	China (SW Sichuan, E Xizang, NW Yunnan)	2500–4000	FOC Vol. 5, p. 344
35	*R. lhasaense* A. J. Li & P. G. Xiao	China (Xizang)	4200–4600	FOC Vol. 5, p. 345

FOC = *Flora of China* (Wu et al. 2003) and FOP = *Flora of Pakistan* (Ali 1980)

comprises of 48 genera and 1200 species at the global scale (Christenhusz and Byng 2016), while 18 genera and 164 species are reported in India (Karthikeyan 2000). However, Mabberley has reported 52 genera and 1200 species from the family at global level (Baas 2017). Also, due to region-political disparity, some studies have reported 43–55 genera for the family (Sanchez et al. 2009). Though cosmopolitan in occurrence, present in almost all ecosystems, it displays more diversity in the North Temperate Zone with the largest genera being *Calligonum* (80 species), *Persicaria* (100 species), *Coccoloba* (120 species), *Rumex* (200 species), and *Eriogonum* (240 species) (Freeman and Reveal 2005). Besides containing some weeds (*Persicaria* sp., *Rumex* sp., etc.), certain species of the Polygonaceae family are also being cultivated for ornamental purposes (Huxley and Griffiths 1999; Uddin et al. 2014).

The systematic position of the family as per the recent two-superkingdom (Prokaryota and Eukaryota), seven-kingdom (Archaea [Archaebacteria] and Bacteria [Eubacteria], and the eukaryotic kingdoms Protozoa, Chromista, Fungi, Plantae, and Animalia) classification proposed by Ruggiero et al. (2015) is as follows:

- **Superkingdom:** Eukaryota
- **Kingdom:** Plantae
- **Subkingdom:** Viridiplantae
- **Infrakingdom:** Streptophyta
- **Superphylum:** Embryophyta
- **Phylum:** Tracheophyta
- **Subphylum:** Spermatophytina
- **Superclass:** Angiospermae
- **Class:** Magnoliopsida
- **Superorder:** Caryophyllanae
- **Order:** Caryophyllales

Botany and Ecology

The Polygonaceae Juss. have long been documented as an isolated family (e.g., Takhtajan 1997; Cronquist 1981; Bentham and Hooker 1880; Meisner 1856) partly distinguished by the presence of leaf ochrea. The demarcation was further supported by molecular phylogenetic studies presenting it as a monophyletic group and sister to the morphologically less-resembling Plumbaginaceae Juss. (Sanchez 2011). Additionally, following different taxonomic rules/criteria for classifying organisms, various taxonomists have given varied classification systems for the family Polygonaceae, as shown in Table 2.2. However, the recent Angiosperm Phylogeny Group (APG)-IV classification system has come up with the following classification for Polygonaceae belonging to the order caryophyllales Juss. ex Bercht. & J. Presl. (Byng et al. 2016).

- **Division:** Angiosperms
- **Clade:** Eudicots
- **Clade:** Core eudicots
- **Clade:** Superasterids
- **Order:** Caryophyllales

In one of the earliest investigations of the family in 1856, Meisner had recognized four subfamilies within it, viz. Brunnichioideae, Eriogonoideae, Polygonoideae, and Symmerioideae which were later reduced to three subfamilies (Coccoloboideae, Eriogonoideae, and Polygonoideae) by other taxonomists including Bentham and Hooker (1880), Dammer (1887), Perdrigeat (1900), and Gross (1912b). With support from earlier reports by Jaretzky (1925), and Haraldson (1978) who dealt with Polygonoideae only, the previous and last review of the Polygonaceae was done in 1993 by Brandbyge who divided it into two subfamilies: Polygonoideae Eaton and Eriogonoideae Arn. The report was mainly based on a set of characteristic morphological features (presence/absence of ocrea and involucre, and monopodial/sympodial growth pattern) and was published in the book *The Families and Genera of Vascular Plants* (Brandbyge 1993). Nonetheless, since then, and based on DNA-led molecular phylogenetic studies, the circumscriptions of these two long-recognized subfamilies have seen some rearrangements (Sanchez and Kron 2008; Sanchez et al. 2009). Both Haraldson (1978) and Brandbyge (1993) divided the subfamilies into different tribes and the genus *Rheum* was kept in the monophyletic tribe Rumiceae (Sanchez et al. 2009). Even though, the genus *Rheum* is composed of nearly 60 species (Sun et al. 2012), only 44 species names are supposed to be the accepted names as per the record available on "The Plant List" web page (http://www.theplantlist.org/1.1/browse/A/Polygonaceae/Rheum/)—accessed on November 1, 2018. The accepted species names available on the webpage are given below and amount to a mere 38.6% of the total 114 scientific plant names of species rank for the genus *Rheum*.

1. *R. nobile* Hook. f. & Thomson
2. *R. tibeticum* Maxim. ex Hook. f.
3. R. wittrockii C.E. Lundstr.

TABLE 2.2

Systematic Position of Polygonaceae Family in Different Classification Systems

Rank	Bentham & Hooker (1862–1883)	Engler & Prantl (1887–1915)	Hutchinson (1973)	Cronquist (1988)	Takhtajan (2009)	Reveal (2012)	Shipunov (2016)
Kingdom	–	Plantae	–	–	–	–	–
Phylum	–	–	Angiospermae	–	Magnoliophyta	–	–
Subphylum	–	–	Dicotyledones	–	–	–	–
Division	–	Embryophyta	Herbaceae	Magnoliophyta	–	–	–
Subdivision	–	Angiospermae	–	–	–	–	–
Class	Dicotyledons	Dicotyledoneae	–	Magnoliopsida	Magnoliopsida	Equisetopsida	Angiospermae
Subclass	Monochlamydeae	Archichlamydeae	–	Caryophyllidae	Caryophyllidae	Magnoliidae	Asteridae
Super order	–	–	–	–	Polygonanae	Caryophyllanae	Caryophyllanae
Order	Curvembryeae	Polygonales	Polygonales	Polygonales	Polygonales	Polygonales	Caryophyllales
Suborder	–	–	–	–	–	–	Plumbaginineae

Botany and Ecology 13

4. *R. australe* D. Don
5. *R. webbianum* Royle
6. *R. globulosum* Gage
7. *R. maculatum* C. Y. Cheng & T. C. Kao
8. *R. compactum* L.
9. *R. alexandrae* Batalin
10. *R. przewalskyi* Losinsk.
11. *R. rhizostachyum* Schrenk
12. *R. spiciforme* Royle
13. *R. rhomboideum* Losinsk.
14. *R. tataricum* L. f.
15. *R. uninerve* Maxim.
16. *R. nanum* Siev. ex Pall.
17. *R. sublanceolatum* C. Y. Cheng & T. C. Kao
18. *R. racemiferum* Maxim.
19. *R. inopinatum* Prain.
20. *R. subacaule* Sam.
21. *R. pumilum* Maxim.
22. *R. delavayi* Franch.
23. *R. yunnanense* Sam.
24. *R. kialense* Franch.
25. *R. tanguticum* Maxim. ex Balf.
26. *R. laciniatum* Prain.
27. *R. hotaoense* C. Y. Cheng & T. C. Kao
28. *R. altaicum* Losinsk.
29. *R. glabricaule* Samuelsson
30. *R. forrestii* Diels
31. *R. likiangense* Sam.
32. *R. lhasaense* A. J. Li & P. G. Xiao
33. *R. acuminatum* Hook. f. & Thomson
34. *R. officinale* Baill.
35. *R. rhabarbamm* L.
36. *R.* x *hybridum* Murray
37. *R. macrocarpum* Losinsk.
38. *R. turkestanicum* Janisch.
39. *R. lucidum* Losinsk.
40. *R. moorcroftianum* Royle
41. *R. reticulatum* Losinsk.
42. *R. palmatum* L.
43. *R. rhaponticum* L.
44. *R. ribes* L.

Diversity in morphological traits of the genus *Rheum*, possibly evolved due to robust selection pressures in arid habitats (Wake et al. 2011), have been broadly used to establish different infra-generic sections in it (Losina-Losinskaya 1936; Kao and Cheng 1975; Li 1998). Earlier, Losina-Losinskaya (1936) recognized nine sections from this genus based on morphology, pollen exine structure, and cpDNA trnL-F region, out of which only five sections were acknowledged by Kao and Chang (1975) who also erected two new sections on the basis of leaf morphology (sect. Acuminata) and inflorescence (sect. Globulosa). Nonetheless, to date, eight sections have been established and acknowledged based on six different types of pollen grains under the genus *Rheum* (Li 1998). The pollen grain types include finely reticulate type (sects; *Rheum* and Palmata); microechinate-foveolate type (sects; *Rheum*, Palmata, Acuminata, Deserticola, Orbicularia, Nobilia, and Spiciformia); microechinate-perforate type (sects: *Rheum*, Palmata, Deserticola, and Spiciformia); rugulate type (sect. Spiciformia); Verrucate-perforate type (sect. Globulosa) and verrucate-rugulate type (sect. Nobilia). Moreover, Yang et al. (2001), and Wang et Al. (2005) has described the pollen morphology in 40 species of genus *Rheum* consisting of all the eight sections (Table 2.3).

2.3 MORPHOLOGY

The family Polygonaceae Juss., also known as the buckwheat family, exhibits inordinate morphological diversity (Ayodele and Olowokudejo 2006). It is cosmopolitan in distribution and found in almost all ecosystems from alpine and tundra to deserts and sand dunes. In fact, some plants occur in rainforests and some are also found in aquatic habitats (Sanchez et al. 2009). The family is monophyletic in origin with the morphological synapomorphies of orthotropous ovules, trigonal achenes, an ocrea, and quincuncial aestivation (Judd et al. 1999). Further, Polygonaceous members are characterized by alternate leaves, regular and perfect flowers, petaloid perianth with no petals, gynoecium tricarpellary, and with one orthotropous and erect ovule containing copious endosperm (Bessey 1915).

The genus *Rheum* consists of erect perennial herbaceous plants with stout and long roots (rhizome). The plant height ranges from 2–8 cm (*R. globulosum*) up to 1–3 m (*R. australe*). Most of the species do have stems but some species are stemless like *R. nanum*, *R. uninerve*, *R. moorcroftianum*, *R. globulosum*, and *R. spiciforme*, etc. The stem is erect, hollow, and glabrous, sulcate, or strigose but solid and warty in *R. ribes* (Ali 1980)—*Flora of Pakistan*, Vol. 205. The leaves are cauline and basal, simple, palmate, or sinuate dentate, mostly basal in rosette with the stem leafless above. However, in some species two to three stem leaves are present, viz. *R. lhasaense*, *R. likiangense*, *R. altaicum*, and *R. forrestii*, etc. The ocrea are usually large, with petiolated leaves; the petiole color is purple-red and yellow, hispidulous in *R. kialense*. The petiole length varies from 5 cm in *R. ribes* to 45 cm in *R. australe*. The leaves are ovate with variations: base cordate, margin entire, or sinuolate. The leaf blade is papilliferous, muricate, or glabrous abaxially, and glabrous, scabrous adaxially. Inflorescence panicle raceme are densely or sparsely branched, with much variation in color which helps in easy and primary recognition of different species. The flowers are small, dense or 1–5 fascicled. The genus is characterized with an absence of petals, though with persistent perianth consisting of six elliptic tepals. Flowers are polygamo-monoecious or bisexual. The androecium consists of nine (six + three) or fewer (seven or eight) purple-red to yellow stamens with variations in anther color. Gynoecium

TABLE 2.3
List of Different Sects of Rheum and the Corresponding Species as Given by Yang et al. (2001) and Wang et al. (2005)

Section	Species
Sect. I *Rheum*	*R. undulatum* L.
	R. webbianum Royle
	R. australe D. Don
	R. likiangense Sam.
	R. forrestii Diels
	R. lhasaense A. J. Li & P. G. Xiao
	R. rhaponticum L.
	R. hotaoense C. Y. Cheng & T. C. Kao
	R. compactum L.
	R. wittrockii Lundstr.
	R. *franzenbachii*
	R. glabricaule Samuelsson
Sect. II Palmata	*R. palmatum* L.
	R. officinale Baill.
	R. tanguticum Maxim. ex Balf.
	R. laciniatum Prain
Sect. III Acuminata	*R. kialense* Franch.
	R. acuminatum Hook. f. & Thomson
	R. maculatum C. Y. Cheng & T. C. Kao
Sect. IV Deserticola	*R. nanum* Siev. ex Pall.
	R. pumilum Maxim.
	R. tibeticum Maxim. ex Hook. f.
	R. uninerve Maxim.
	R. delavayi Franch.
	R. sublanceolatum C. Y. Cheng & T. C. Kao
	R. racemiferum Maxim.
	R. inopinatum Prain
Sect. V Orbicularia	*R. tataricum* L. f.
Sect. VI Spiciformia	*R. moorcroftianum* Royle
	R. rhizostachyum Schrenk
	R. spiciforme Royle
	R. reticulatum Losinsk.
	R. przewalskyi Losinsk.
	R. rhomboideum Losinsk.
Sect. VII Globulosa	*R. globulosum* Gage
Sect. VIII Nobilia	*R. nobile* Hook. f. & Thomson
	R. alexandrae Batalin

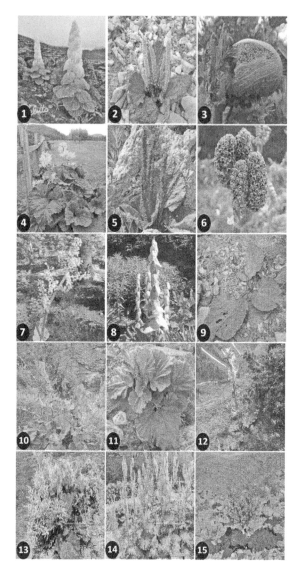

FIGURE 2.1 Representative species of genus Rheum in natural habitats. 1- *R. nobile*, 2- *R. spiciforme*, 3- *R. tanguticum*, 4- *R. rhabarbarum*, 5- *R. rhizostachyum*, 6- *R. kialense*, 7- *R. palmatum*, 8- *R. alexandrae*, 9- *R. tibeticum*, 10- *R. webbianum*, 11- *R. moorcroftianum*, 12- *R. australe*, 13- *R. officinale*, 14- *R. altaicum*, 15- *R. palaestinum*

consists of large and inflated stigma, recurved, three short and cylindrical styles, rhomboid to ovoid ovary. The fruit is achene oblong to ovoid with brown, and with ovoid to ellipsoid trigonous winged seeds (Wu et al. 2003)—*Flora of China*; Vol. 5, pp. 341–350. A detailed sketch of morphological characteristics of different species of the genus *Rheum* is outlined in Table 2.4 and some images from representative species of the genus are provided in Figure 2.1.

TABLE 2.4
Detailed Description of Morphometric Characteristics of Different Species of the Genus *Rheum* as Reported in *Flora of China* and *Flora of Pakistan*

Character/Species	*R. acuminatum*	*R. nobile*
Growth habit	Erect	Erect
Plant height	50–80 cm	1–2 m
Stem- present/absent	Present	Present
Stem- solid/hollow	Hollow, glabrous	Glabrous
Basal leaves	1 to 3	In a rosette
Blade abaxial side	NA	NA
Blade adaxial side	NA	NA
Leaf shape	Ovate	Lanceolate
Leaf margin	Entire	NA
Leaf base	Cordate	Cordate-ovate
Leaf apex	NA	NA
Inflorescence type	Panicle	Panicle
Inflorescence color	Purple-red	Yellow-green
Inflorescence branching	NA	5–8 branched
Flowers	10 fascicled	5–9 fascicled
Perianth	Tepals 6	Tepals 6
Androecium	Stamens 8–9	Stamens 8 (9)
Anther	Black-purple	Compressed, oblong-elliptic
Ovary	Rhomboid-ellipsoid	Ovoid
Fruit	Oblong-ovoid	Ovoid
Seeds	Ovoid, brown	NA
Reference	FOC Vol. 5, p. 346	FOC Vol. 5, p. 350

R. ribes	*R. tibeticum*	*R. moorcroftianum*
Erect	Erect	Erect
1 m	NA	NA
Present	Present	Absent
Solid, warty	NA	NA
Only basal	All	NA
Verrucose	Papillose	NA
Glabrous	Glabrous-papillose	NA
Obtuse	NA	Broadly ovate
NA	Entire or crenulate	Entire
Cordate	Orbicular	NA
Undulate	NA	NA
NA	Leafless panicle	Pedunculate
NA	NA	NA
NA	NA	NA
NA	NA	NA

(Continued)

TABLE 2.4 (CONTINUED)
Detailed Description of Morphometric Characteristics of Different Species of the Genus *Rheum* as Reported in *Flora of China* and *Flora of Pakistan*

Tepals 6	Tepals 6	Tepals 6
Stamens 8–9	Stamens 8–9	Stamens 9 or fewer
NA	NA	NA
NA	NA	NA
Ovate-oblong	Orbicular	Ellipsoid or oblong
Dull brown	NA	NA
FOP Vol. 205	FOP Vol. 205	FOC Vol. 5, p. 349
R. wittrockii	***R. australe***	***R. webbianum***
Erect	Erect	Erect
1 m	1–3 m	0.5–1.5 m
Present	Present	Present
NA	Glabrous or pubescent	Hollow, finely sulcate, glabrous
NA	NA	NA
Hairy below	Papillose	NA
Glabrous	Scabrous	NA
NA	Ovate-elliptic	Ovate
NA	Entire	Sinuolate
Cordate	Cordate	Cordate
Acute	Sinuolate	Obtuse
Panicle	Panicle	Panicle
NA	Dark purple	Yellow-white
NA	Fastigiately branched	1- or 2-branched
NA	NA	Flowers small
Tepals 6	Tepals 6	Tepals 6
Stamens 8–9	Stamens 9 or fewer	Stamens 8–9
NA	NA	NA
NA	Rhomboid-obovoid	NA
Oblong-orbicular	Oblate and muricate	Ellipsoid or orbicular
Nut brown	Ovoid-ellipsoid	Ovoid-ellipsoid
FOC Vol. 5, p. 344	FOP Vol. 205	FOC Vol. 5, p. 343
R. globulosum	***R. spiciforme***	***R. reticulatum***
Erect	Erect	Erect
2–8 cm	Short, stout	Short
Absent	Absent	Present
NA	NA	NA
1, rarely 2	Subterete	NA
Glabrous or muricate	Glabrous	Papilliferous

(Continued)

Botany and Ecology

TABLE 2.4 (CONTINUED)
Detailed Description of Morphometric Characteristics of Different Species of the Genus *Rheum* as Reported in *Flora of China* and *Flora of Pakistan*

Leathery	Papilliferous	Glabrous
Reniform-orbicular	Ovate	Ovate to triangular-ovate
Entire	Entire	Slightly sinuolate
Cordate	Rounded or Sub-cordate	Rounded or Sub-cordate
Obtuse	Obtuse	Acute
Panicle headlike	Panicle spiciform	Panicle spiciform
Light green with white margin	Light green	Yellow-white
Short	3 mm, slender	Short, 1.5–2 mm
Flowers dense	NA	Flowers dense
Tepals 6	Tepals 6	Tepals 6
Stamens 8–9	Stamens 9 or fewer	Stamens 8–9
NA	Anthers yellow	NA
Rhomboid	Obovoid	Obovoid-ellipsoid
Ovoid	Fruit oblong-ellipsoid	Broadly ovoid
NA	NA	Ovoid
FOC Vol. 5, p. 350	FOC Vol. 5, p. 349	FOC Vol. 5, p. 350
R. maculatum	*R. compactum*	*R. alexandrae*
Erect	Erect	Erect
50–80 cm	1 m	40–80 cm
Present	Present	Present
Hollow, slender	Hollow, glabrous	Hollow, glabrous
Many	Petiole subterete	4 to 6
White hairs	Pubescent	Glabrous
Glabrous	NA	Glabrous
Reniform-cordate	Ovate-cordate	Ovate to ovate-elliptic
Entire	Entire	NA
Cordate	Cordate	NA
Acuminate	Obtuse	NA
Panicle terminal	Fascicled	Panicle
Light red	Yellow	Green
Few branched	Densely branched	2- or 3-branched
Flowers small	Flowers fascicled	Fascicled, small
Tepals 6	Tepals 6	Tepals 6
Stamens 8–9	Stamens 9 or fewer	Stamens 7–9
Anthers purple, sub-globose	NA	NA
NA	NA	Rhomboid-obovoid

(Continued)

20 — Genus *Rheum*: A Global Perspective

TABLE 2.4 (CONTINUED)
Detailed Description of Morphometric Characteristics of Different Species of the Genus *Rheum* as Reported in *Flora of China* and *Flora of Pakistan*

Broadly ellipsoid	Oblong-ellipsoid	Rhomboid-ellipsoid
NA	Ovoid	NA
FOC Vol. 5, p. 346	FOC Vol. 5, p. 345	FOC Vol. 5, p. 350
R. przewalskyi	*R. rhizostachyum*	*R. rhomboideum*
Erect	Erect	Erect
Short, stout	30 cm	Dwarf, procumbent
Absent	Present	Absent
NA	NA	NA
2 to 4	NA	All
Glabrous or papilliferous	Densely papilliferous	Densely papilliferous
Glabrous	NA	Glabrous
Ovate or rhombic-ovate	Ovate	Rhombic or elliptic
NA	Entire	Entire
NA	Cordate or rounded	Cuneate
NA	Obtuse	Obtuse or obtusely acute
NA	Panicles 2–5, spiciform	Panicle spiciform
Yellow-white	Yellow-white	Purple-red
NA	NA	NA
NA	NA	NA
Tepals 6	Tepals 6	Tepals 6
Stamens 9 or fewer	Stamens 9 or fewer	Stamens 9 or fewer
NA	NA	NA
Ellipsoid	NA	NA
Ovoid	Ovoid or ellipsoid-ovoid	Broadly cuneate
NA	Ovoid	Ovoid
FOC Vol. 5, p. 349	FOC Vol. 5, p. 350	FOC Vol. 5, p. 349
R. tataricum	*R. uninerve*	*R. nanum*
Erect	Erect	Erect
35–50 cm	15–30 cm	20–35 cm
Present	Absent	Absent
Hollow, glabrous	NA	NA
Procumbent, large	2 to 4	2 to 4
Glabrous or papilliferous	NA	Glabrous
NA	NA	Tuberculate
Cordate-orbicular	NA	NA
Serrulate	Sinuolate	Entire

(Continued)

Botany and Ecology

TABLE 2.4 (CONTINUED)
Detailed Description of Morphometric Characteristics of Different Species of the Genus *Rheum* as Reported in *Flora of China* and *Flora of Pakistan*

Cordate	Rounded or cuneate	Rounded or Sub-cordate
Obtuse	Obtuse-acute	Rounded
Panicle	Panicle narrow	Panicle broad
Yellow-white	Purple-red	Yellow-white
3-branched	1- or 2-branched	Branched at middle
1- or 2-fascicled	2-4 fascicled	Densely fascicled
Tepals 6	Tepals 6	Tepals 6
Stamens 9 or fewer	Stamens 8 or 9	Stamens 9 or fewer
NA	NA	NA
Triangular-ovoid	Rhomboid-ovoid	Rhomboid-ellipsoid
Purple-red, ovoid	Oblong-ellipsoid	Reniform
Dark brown, ovoid	Dark brown ovoid	Ovoid
FOC Vol. 5, p. 349	FOC Vol. 5, p. 348	FOC Vol. 5, p. 348
R. sublanceolatum	*R. racemiferum*	*R. inopinatum*
Erect	Erect	Erect
30–55 cm	50–70 cm	20–35 cm
Present	Present	Present
Hollow	Hollow	NA
3 to 5	2 to 5	3 to 5 in a rosette
NA	NA	NA
NA	NA	NA
Ovate or lanceolate	Cordate or ovate	Triangular-ovate
Entire or rarely sinuolate	Sinuolate	Sinuolate
Acute	Sub-cordate	Cordate or truncate
NA	Obtuse	Obtuse
Panicle narrow	Panicle	Panicle terminal
NA	NA	Yellow-white
Once branched	Once branched	Once
4- or 5-fascicled	Fascicled	NA
Tepals 6	Tepals 6	Tepals 6
Stamens 9 or fewer	Stamens 9 or fewer	Stamens 9 or fewer
NA	NA	NA
NA	NA	Ovoid
Ovoid-ellipsoid	Ellipsoid to oblong-ellipsoid	Orbicular
Brown, ovoid	Dark brown, ovoid-ellipsoid	Ovoid
FOC Vol. 5, p. 347	FOC Vol. 5, p. 347	FOC Vol. 5, p. 347

(Continued)

TABLE 2.4 (CONTINUED)
Detailed Description of Morphometric Characteristics of Different Species of the Genus *Rheum* as Reported in *Flora of China* and *Flora of Pakistan*

R. subacaule	*R. pumilum*	*R. delavayi*
Erect	Erect	Erect
15–20 cm	10–25 cm	15–28 cm
Present	Present	Present
Densely pubescent	Finely striped, slender, pilose	Pilose
3 or 4	2 or 3	2 to 4
Hispidulous	Pilose	Hispidulous
Glabrous	Glabrous	Glabrous
NA	Ovate-elliptic	Oblong-elliptic or ovate-elliptic
Entire	Entire	Entire
Base cordate	Cordate	Sub-cordate
Acuminate	Obtuse	Obtuse
Panicles	Panicle narrow	Panicle narrow
Purple-red	Purple-red	Purple
2-4 from base	Sparsely branched	Branched once
3-5-fascicled	2- or 3-fascicled	3- or 4-fascicled
Tepals 6	Tepals 6	Tepals 6
Stamens 9 or less	Stamens 9 or less	Stamens 9 or less
NA	NA	Purple
Ellipsoid	Ellipsoid	Elliptic to subglobate
Cordate	Triangular-ovoid	Cordate-orbicular
Triangular-ovoid; red-brown	Ovoid	Ovoid
FOC Vol. 5, p.348	FOC Vol. 5, p. 347	FOC Vol. 5, p. 348

R. yunnanense	*R. kialense*	*R. tanguticum*
Erect	Erect	Erect
30–60 cm	25–55 cm	(0.6–)1.5–2 m
Present	Present	Present
NA	Hollow	Hollow
1 to 3	1 to 3	Few
Pubescent	Sparsely hispidulous	Pubescent
Glabrous	NA	Papilliferous or muricate
Reniform-cordate	Ovate-cordate	Orbicular or broadly ovate
Entire	Entire	NA
Cordate	Cordate	Sub-cordate
Acuminate	Slightly acuminate	Narrowly acute
Panicle	Panicle	Panicle large
Purple	White-green	Purple-red

(Continued)

Botany and Ecology

TABLE 2.4 (CONTINUED)
Detailed Description of Morphometric Characteristics of Different Species of the Genus *Rheum* as Reported in *Flora of China* and *Flora of Pakistan*

1- or 2-branched	Few branched	Branches connivant
1-3(or 4)-fascicled	2-5-fascicled	Small
Tepals 6	Tepals 6	Tepals 6
Stamens 9 or less	Stamens 9 or less	Stamens 9 or less
Purple-red	Purple-red	NA
Narrowly ellipsoid	Rhomboid-ellipsoid	Broadly ovoid
Purple-red, ellipsoid or ovoid	Ovoid or sub-ovoid	Oblong-ovoid
NA	Narrowly ovoid; yellow-brown	Black, ovoid
FOC Vol. 5, p. 346	FOC Vol. 5, p. 347	FOC Vol. 5, p. 345
R. laciniatum	*R. hotaoense*	*R. altaicum*
Erect	Erect	Erect
1 m	0.8--1.5 m	50–100 cm
Present	Present	Present
NA	Glabrous, muricate	Hollow
NA	NA	NA
Muricate	NA	Papilliferous or pubescent
NA	NA	Glabrous
Subovate	Ovate-cordate	Ovate-cordate or triangular-ovate
NA	Sinuolate	Slightly sinuolate
NA	Cordate	Cordate
Acute	Obtuse or acute	Obtuse
Panicle triangular	Panicle large	Panicle narrow
Yellow-white	Light green with white margin	Yellow-white
NA	More than 2-branched	NA
NA	Flowers large	Small, 4–7-fascicled
Tepals 6	Tepals 6	Tepals 6
Stamens 9 or less	Stamens 9 or less	Stamens 9 or less
NA	NA	NA
Sub-ovoid	Broadly ellipsoid	Oblong-ellipsoid
NA	Orbicular	Black-brown, broadly ovoid
NA	Broadly ovoid	NA
FOC Vol. 5, p. 346	FOC Vol. 5, p. 344	FOC Vol. 5, p. 344
R. glabricaule	*R. forrestii*	*R. likiangense*
Erect	Erect	Erect
1 m	60–80 cm	40–70(–90) cm
Present	Present	Present

(Continued)

24 Genus *Rheum*: A Global Perspective

TABLE 2.4 (CONTINUED)
Detailed Description of Morphometric Characteristics of Different Species of the Genus *Rheum* as Reported in *Flora of China* and *Flora of Pakistan*

NA	Hollow	NA
All	3 to 5	2 to 4
Pubescent	Densely hispid	Glabrous
Glabrous	Pubescent or glabrous	NA
Cordate-ovate	Broadly ovate or ovate	NA
Entire	Entire	Entire
NA	Cordate	Cordate
Acuminate	Obtuse	Obtuse or acute
Panicle narrow	Panicle	Panicle
Green with purple margin	Yellow-green	White-green
Sparsely branched	Branched from middle	1- or 2-branched
NA	Densely fascicled	Fascicled
Tepals 6	Tepals 6	Tepals 6
Stamens 9 or less	Stamens 9 or less	Stamens 9 or less
Purple	Purple-red	NA
Purple, oblong-ovoid	Slightly inflated	Rhomboid-circular
	Broadly ellipsoid or orbicular	Ovoid
Ovoid	Yellow-brown, ovoid-ellipsoid	Ovoid
FOC Vol. 5, p. 345	FOC Vol. 5, p. 344	FOC Vol. 5, p. 344

R. ihasaense
Erect
30–70 cm
Present
Glabrous, or pubescent
NA
Shortly hispid
Glabrous
Narrowly triangular or triangular-ovate
Slightly sinuolate
Cordate
Obtuse or acute
Panicles narrow
Light green with purple margin
2-branched
NA
Tepals 6
Stamens 9 or less

(Continued)

Botany and Ecology 25

TABLE 2.4 (CONTINUED)
Detailed Description of Morphometric Characteristics of Different Species of the Genus *Rheum* as Reported in *Flora of China* and *Flora of Pakistan*

NA
NA
Spherical
Orbicular or ovoid
FOC Vol. 5, p. 345

2.4 SYNONYMS AND VERNACULAR NAMES

Rheum is commonly known as rhubarb and has been used as a medicinal plant since ancient times in the Chinese system of medicine. With distribution centers in western and north-western China, most of its species are still confined to China. It is called "Ta huang" and "chun tza" in traditional Chinese and Tibetan medical systems, respectively (Xiao et al. 1984). The people of China call the genus *Rheum* "huang" and had thus named its different species with a "huang" suffix. As most of the species are mainly confined to China, the vernacular names for them are also in the Chinese or Tibetan languages. A few species are also found in India, Pakistan, Nepal, and other Asian countries and are accordingly named locally in Hindi, Urdu, Nepalese, and other languages. Further, different species of the genus *Rheum* are named according to their use, such as "ornamental rhubarb," or the country of endemism such as "Turkish rhubarb," etc. A detailed description of the vernacular names of different species of the genus Rheum is provided in Table 2.5.

Further, as per the web record (accessed on November 18, 2018) available on "The Plant List" web page (http://www.theplantlist.org/1.1/browse/A/Polygonaceae/Rheum/), various species of *Rheum* have their synonyms as well; for example, *R. emodi* is the synonym for *R. australe*, *R. delavayi* for *R. strictum*, etc. Table 2.6 gives an update on the status of synonyms used to date for the enlisted species of the genus *Rheum*.

2.5 POLLEN STUDIES

Palynology, the study of various phases of pollen and spores, has added a new dimension to the taxonomy of angiosperms both at generic and specific levels. Pollen morphology (aperture, exine ornamentation/strata, shape, and size) is indeed the expression of genome with a unique position in deciphering remarkable information from the miniature structural components of the plant itself. The significance of pollen sculpturing is best evidenced in taxonomy and phylogeny

TABLE 2.5
List of Various Plant Species of *Rheum* with Their Vernacular Names

S. No.	Scientific Name	Vernacular Name (Language or Locality)	Reference
1	R. nobile	Ta huang (Chinese), chulthi amilo (Nepalese)	FOC Vol. 5, p. 350
2	R. tibeticum	Xi zang da huang (Chinese), lat-chu (Ladakhi)	FOC Vol. 5, p, 348
3	R. wittrockii	Tian shan da huang (Chinese)	FOC Vol. 5, p. 344
4	R. australe	Zang bian da huang (Chinese), revand-chini (Hindi), padamchal (Nepalese), chotial (Pakistan)	FOC Vol. 5, p. 343, FOP Vol. V, p. 205, (Agarwal et al. 2000), (Hamayun et al. 2007), (Gupta et al. 2014)
5	R. webbianum	Xu mi da huang (Chinese), rhubarb (English), chotal (Gilgit area)	FOC Vol. 5, p. 343, FOP Vol. V, p. 205
6	R. globulosum	Tou xu da huang (Chinese)	FOC Vol. 5, p. 350
7	R. maculatum	Ban jing da huang (Chinese)	FOC Vol. 5, p. 346
8	R. compactum	Ban jing da huang (Chinese)	FOC Vol. 5, p. 345
9	R. alexandrae	Shui huang (Chinese)	FOC Vol. 5. P. 350
10	R. przewalskyi	Qi sui da huang (Chinese)	FOC Vol. 5, p. 349
11	R. rhizostachyum	Zhi sui da huang (Chinese)	FOC Vol. 5, p. 350
12	R. spiceforme	Sui xu da huang (Chinese), rewand (Pakistan)	FOC Vol. 5, p. 349, FOP Vol. V, p. 205
13	R. rhomboideum	Ling ye da huang (Chinese)	FOC Vol. 5, p. 349
14	R. tataricum	Yuan ye da huang (Chinese), tatar rhubarb (English)	FOC Vol. 5, p. 349, (Samappito et al. 2003)
15	R. uninerve	Dan mai da huang (Chinese)	FOC Vol. 5, p. 348
16	R. nanum	Ai da huang (Chinese)	FOC Vol. 5, p. 348
17	R. sublanceolatum	Zhai ye da huang (Chinese)	FOC Vol. 5, p. 347
18	R. racemiferum	Zong xu da huang (Chinese)	FOC Vol. 5, p. 347
19	R. inopinatum	Hong mai da huang (Chinese)	FOC Vol. 5, p. 347
20	R. subacaule	Chui zhi da huang (Chinese)	FOC Vol. 5, p. 348
21	R. pumilum	Xiao da huang (Chinese)	FOC Vol. 5, p. 347
22	R. delavayi	Dian bian da huang (Chinese)	FOC Vol. 5, p. 348
23	R. yunnanens	Yun nan da huang (Chinese)	FOC Vol. 5, p. 346
24	R. kialense	Shu zhi da huang (Chinese)	FOC Vol. 5, p. 347
25	R. tanguticum	Ji zhua da huang (Chinese), tangut Rheum (English)	FOC Vol. 5, p. 345, (Jin et al. 2007), (Wang and Ren 2009)
26	R. laciniatum	Tiao lie da huang (Chinese)	FOC Vol. 5, p. 346
27	R. hotaoense	He tao da huang (Chinese)	FOC Vol. 5, p. 344
28	R. altaicum	A er tai da huang (Chinese)	FOC Vol. 5, p. 344
29	R. glabricaule	Guang jing da huang (Chinese)	FOC Vol. 5, p. 345
30	R. forrestii	Niu wei qi (Chinese)	FOC Vol. 5, p. 344
31	R. likiangense	Li jiang da huang (Chinese)	FOC Vol. 5, p. 344

(Continued)

Botany and Ecology

TABLE 2.5 (CONTINUED)
List of Various Plant Species of *Rheum* with Their Vernacular Names

S. No.	Scientific Name	Vernacular Name (Language or Locality)	Reference
32	*R. lhasaense*	La sa da huang (Chinese)	FOC Vol. 5, p. 345
33	*R. acuminatum*	Xin ye da huang (Chinese)	FOC Vol. 5, p. 346
34	*R. officinale*	Yao yong da huang (Chinese), chun tza (Tibetan)	FOC Vol. 5, p. 345, (Xiao et al. 1984), (Huang et al. 2007)
35	*R. palmatum*	Zhang ye da huang (Chinese), Chun tza (Tibetan), Chinese rhubarb, ornamental rhubarb, Turkish rhubarb (English), kelembak (Indonesia)	FOC Vol. 5, p. 345, (Xiao et al. 1984), (Kubo et al. 1992), (Huang et al. 2007), (Miraj 2016)
36	*R. rhabarbamm*	Bo ye da huang (Chinese), garden rhubarb, pie-plant, wine-plant (English)	FOC Vol. 5, p. 343, FNA Vol. 5
37	*R. moorcroftianum*	Luan guo da huang (Chinese)	FOC Vol. 5 p. 349
38	*R. reticulatum*	Wang mai da huang (Chinese)	FOC Vol. 5, p. 350
39	*R. rhaponticum*	Rilski raven (Bulgaria)	FOP Vol. V, p. 205
40	*R. ribes*	Iskin, usgun, ucgun or rivas (Iran), rawash (Pakistani)	FOP Vol. V, p. 205,

* FOC = *Flora of China*, FON = *Flora of Nepal*, FOP = *Flora of Pakistan*, FNA = *Flora of North America*

of the angiosperms at species and even varietal levels. The advances in the techniques of microscopy, particularly scanning electron microscopes (SEMs) during the 1970s, have further refined the field of palynology by acting as an ideal source for relative studies of pollen ornamentation (Ferguson 1985; Wodehouse 1935). The study on the morphological aspects of the polygonaceous pollen grains was earlier made by Eranz and Gross (Gross 1912a), with substantial contributions later made by researchers like Wodehouse (1931), Hedberg (1945), Erdtman (1969, 1952), Nowicke and Skvarla (1977), and Mondal (1997). The latter studied 156 species from 22 genera from native and foreign taxa wherein he used an interdisciplinary botanical approach vis-à-vis palynology to resolve the taxonomic disputes within Polygonaceae and also the inter-familial controversies. Further, the Polygonaceae family from the Kashmir Himalayas has been thoroughly studied by Munshi and Javeid wherein they examined pollen ornamentations of about 50 species (Munshi and Javeid 1986).

The Polygonaceae family is a noticeably eurypalynous family with an extensive range of pollen stratifications. Though small, pollen grains offer a great taxonomic significance in families like Polygonaceae which are highly multipalynous. Additionally, the inter-generic/specific variations within the external

TABLE 2.6
List of Various Plant Species of *Rheum* with Their Synonyms

S. No.	Accepted Name	Synonym	Reference
1	R. nobile	NA	The Plant List, record Tropicos -26001862
2	R. ribes	NA	The Plant List, record Tropicos -26002461
3	R. tibeticum	NA	The Plant List, record Tropicos -26000185
4	R. wittrockii	NA	The Plant List, record Tropicos -26002468
5	R. australe	R. emodi Wall. ex Meisn.	The Plant List, record Tropicos -26001861
6	R. webbianum	NA	The Plant List, record Tropicos -26002590
7	R. globulosum	NA	The Plant List, record Tropicos -50000768
8	R. maculatum	NA	The Plant List, record Tropicos -50136395
9	R. compactum	R. nutans Pall., R. orientale Losinsk.	The Plant List, record Tropicos -26001336
10	R. alexandrae	NA	The Plant List, record Tropicos -50136523
11	R. przewalskyi	NA	The Plant List, record Tropicos -50136495
12	R. rhizostachyum	R. aplostachyum Kar. & Kir	The Plant List, record Tropicos -26002460
13	R. spiciforme	R. scaberrimum Lingelsh.	The Plant List, record Tropicos -26002463
14	R. rhomboideum	NA	The Plant List, record Tropicos -50136486
15	R. tataricum	R. caspicum Pall. R. songaricum Schrenk	The Plant List, record kew-2424199
16	R. uninerve	NA	The Plant List, record Tropicos 50136442
17	R. nanum	R. cruentum Siev. ex Pall. R. leucorrhizum Pall.	The Plant List, record Tropicos -26002454
18	R. sublanceolatum	NA	The Plant List, record Tropicos -50136426
19	R. racemiferum	NA	The Plant List, record Tropicos -50136412
20	R. inopinatum	NA	The Plant List, record Tropicos -50136429
21	R. subacaule	NA	The Plant List, record Tropicos -50136440
22	R. pumilum	NA	The Plant List, record Tropicos -50136431
23	R. delavayi	R. strictum Franch.	The Plant List, record Tropicos -50136437
24	R. yunnanense	NA	The Plant List, record Tropicos -50136399
25	R. kialense	R. micranthum Sam.	The Plant List, record Tropicos -50136403
26	R. tanguticum	R. palmatum subsp. dissectum Stapf	The Plant List, record Kew-2424196
27	R. laciniatum	NA	The Plant List, record Tropicos -50000766
28	R. hotaoense	NA	The Plant List, record Tropicos -50136336
29	R. altaicum	R. rhaponticum Herder	The Plant List, record Tropicos -26002441
30	R. glabricaule	NA	The Plant List, record Tropicos -50136364
31	R. forrestii	NA	The Plant List, record Tropicos -26002814
32	R. likiangense	R. ovatum C. Y. Cheng & T. C. Kao	The Plant List, record Tropicos -50136351
33	R. lhasaense	NA	The Plant List, record Tropicos -50136356

(Continued)

Botany and Ecology

TABLE 2.6 (CONTINUED)
List of Various Plant Species of *Rheum* with Their Synonyms

S. No.	Accepted Name	Synonym	Reference
34	*R. moorcroftianum*	NA	The Plant List, record Tropicos -26002354
35	*R. acuminatum*	*R. orientalixizangense* Y. K. Yang, J. K. Wu & Gasang.	The Plant List, record Tropicos -50136394
36	*R. officinale*	NA	The Plant List, record kew-2425521
37	*R. rhaponticum*	*Rhabarbarum rhaponticum* Moench, *R. esculentum* Salisb., *R. rotundatum* Stokes, *R. sibiricum* Pall.	The Plant List, record Kew-2425551
38	*R. reticulatum*	NA	The Plant List, record Tropicos -26002459
39	*R. palmatum*	*R. potaninii* Losinsk., *R. qinlingense* Y. K. Yang, D. K. Zhang & J. K. Wu, *Rhabarbarum palmatum* Moench	The Plant List, record Kew-2425567
40	*R. palaestinum*	NA	The Plant List, record Kew-2425566
41	*R. rhabarbarum*	*R. franzenbachii* Münter, *R. franzenbachii* var. *mongolicum* Münter, *R. undulalum* L., *R. undulatum* L., *R. undulatum* var. *longifolium* C. Y. Cheng & T. C. Kao	The Plant List, record Kew-2425570
42	*R. x hybridum*	NA	The Plant List, record Kew-2425537
43	*R. macrocarpum*	*R. ferganense* Titov, *R. lobatum* Litv. ex Losinsk., *R. nuratavicum* Titov, *R. plicatum* Losinsk., *R. vvedenskyi* Sumner, *R. zergericum* Titov	The Plant List, record Tropicos -26002450
44	*R. turkestanicum*	*R. megalophyllum* Sumner, *R. renifolium* Sumner, *R. rupestre* Litv. ex Losinsk., *R. turanicum* Litv.	The Plant List, record Tropicos -26002466

30 Genus *Rheum*: A Global Perspective

ornamentations of these pollen grains are suggestive of the genetic variability within a group and may even lead to and explain the events of speciation.

As discussed in the previous section, nine sections were earlier recognized (Losina-Losinskaya 1936) within the genus *Rheum* which were later finalized to eight sections based on the pollen ornamentations by Li (1998). For phylogenetic and evolutionary analyses, pollen morphology is considered as important evidence (Erdtman 1952; Walker 1974). Earlier, only *R. delavayi* was reported from genus *Rheum* (Nowicke and Skvarla 1977). Later, another species, *R. officinale* was described by Leeuwen et al. (1988) and Wang et al. (1995) which was followed by pollen data of *R. palmatum* by Zhang and Zhou (1998). In an effort toward understanding the genus *Rheum* on phylogenetic and evolutionary levels vis-à-vis its geographical/ecological distribution in north-western China, Yang et al. (2001) examined the pollen morphology of 40 species of the genus and found that medicinally important *R. officinale, R. palmatum,* and *R. tanguticurn* can be palynologically distinguished which were otherwise morphologically and anatomically interwoven (Li and Zhang 1983). While projecting *Rheum* as a natural group with common ancestry, they further proposed a tentative evolutionary trend for different pollen types (Table 2.7) based on their morphological variations within the genus *Rheum*. Owing to its widespread existence within the genus, the microechinate-foveolate/perforate type is considered as the most primitive type, followed by the finely reticulate type, rugulate type, verrucate-perforate type, and the most advanced verrucate-rugulate type. The primitive pollen types were shown to appear in species occurring at lower altitudes while the advanced types are found in high-altitude-growing species of *Rheum*. Additionally, due to the occurrence of more than one pollen type within the same section, they also anticipated the possible existence of parallel evolution of pollen types in the genus *Rheum* which was further correlated with geographical/ecological parameters. The detailed description of different pollen types in various species of genus *Rheum* are given in Table 2.8.

TABLE 2.7
Characteristics of Different Pollen Types

Pollen type	Description
A	Microechinate-perforate type: Tectum densely perforated with smooth ornamentation
B	Finely reticulate type: Micro-reticulate with microechinate ornamentation
C	Microechinate-foveolate type: Microechinate tectum smooth sparsely foveolate/perforate with even to uneven ornamentation
D	Rugulate type: Indistinctly regulate with low relief ornamentation
E	Verrucate-perforate type
F	Verrucate-rugulate type: High relief, verrucate ornamentation and distinctly rugulate

TABLE 2.8

Detailed Description of Pollen Morphology as Observed in Various Species of Genus _Rheum_

Section	Species	Pollen Type	Pollen Shape	Pollen Size (P x E μm)*	Method	Reference
Sect. I _Rheum_	R. undulatum	C	Subspheroidal-prolate	37.2 × 31.1	LS, SEM	(Yang et al. 2001;
	R. webbianum	C	Subspheroidal	25.3 × 25.2	LS, SEM	Wang et al. 2005)
	R. australe	C	Subspheroidal	26.9 × 25.5	LS, SEM	
	R. likiangense	C	Subspheroidal	28 × 28.9	LS, SEM	
	R. forrestii	C	Subspheroidal	31.6 × 29	LS, SEM	
	R. lhasaense	C	Subspheroidal	33.1 × 32.5	LS, SEM	
	R. hotaoense	A	Prolate	40.8 × 32.6	LS, SEM	
	R. compactum	A	Subspheroidal	28.2 × 26.8	LS, SEM	
	R. wittrockii	A	Subspheroidal-ellipsoidal	30.7 × 31.4	LS, SEM	
	R. franzenbachii	A	Subspheroidal	32.0 × 32.8	LS, SEM	(Yang et al. 2001)
	R. glabricaule	A	Subspheroidal	30.5 × 26.7	LS, SEM	
Sect. II Palmata	R. palmatum	A	Subspheroidal	31.4 × 29.3	LS, SEM	(Yang et al. 2001;
	R. officinale	B	Subspheroidal	30.6 × 28.7	LS, SEM	Wang et al. 2005)
	R. tanguticum	C	Subspheroidal-prolate	33.8 × 22.9	LS, SEM	
	R. laciniatum	B	Subspheroidal	29.9 × 28.3	LS, SEM	(Yang et al. 2001)
Sect. III Acuminata	R. kialense	C	Subspheroidal-prolate	30.9 × 25.3	LS, SEM	(Yang et al. 2001; Wang et al. 2005)
	R. acuminatum	C	Subspheroidal	31.3 × 29.8	LS, SEM	(Yang et al. 2001)
	R. maculatum	C	Subspheroidal-prolate	32.4 × 21.5	LS, SEM	

(Continued)

TABLE 2.8 (CONTINUED)
Detailed Description of Pollen Morphology as Observed in Various Species of Genus *Rheum*

Section	Species	Pollen Type	Pollen Shape	Pollen Size (P x E µm)*	Method	Reference
Sect. IV Deserticola	*R. nanum*	A	Subspheroidal	29.4 × 30.2	LS, SEM	(Yang et al. 2001;
	R. pumilum	A	Subspheroidal	31.1 × 27.8	LS, SEM	Wang et al. 2005)
	R. tibeticum	C	Subspheroidal	28.2 × 25.7	LS, SEM	
	R. uninerve	C	Subspheroidal-prolate	34.2 × 29.5	LS, SEM	(Yang et al. 2001)
	R. delavayi	C	Subspheroidal-prolate	30.6 × 23.8	LS, SEM	
	R. sublanceolatum	C	Prolate	34.1 × 24.7	LS, SEM	
	R. racemiferum	C	Subspheroidal	31.7 × 30.9	LS, SEM	
	R. inopinatum	A	Prolate	34.3 × 28.1	LS, SEM	
Sect. V Orbicularia	*R. tataricum*	C	Prolate	35.2 × 31.1	LS, SEM	
Sect. VI Spiciformia	*R. moorcroftianum*	A	Subspheroidal	28.6 × 27.38	LS, SEM	(Yang et al. 2001;
	R. rhizostachyum	A	Subspheroidal-prolate	28.2 × 26.4	LS, SEM	Wang et al. 2005)
	R. spiciforme	C	Subspheroidal-prolate	28.1 × 28.6	LS, SEM	
	R. reticulatum	C	Subspheroidal-prolate	28.9 × 27.8	LS, SEM	
	R. przewalskyi	D	Subspheroidal-prolate	29.6 × 26.9	LS, SEM	
	R. rhomboideum	C	Subspheroidal-prolate	27.1 × 28.2	LS, SEM	(Yang et al. 2001)
Sect. VII Globulosa	*R. globulosum*	E	Subspheroidal	32.3 × 30.3	LS, SEM	
Sect. VIII Nobilia	*R. nobile*	F	Subspheroidal	23.3 × 24.5	LS, SEM	(Yang et al. 2001; Wang et al. 2005)
	R. alexandrae	C	Prolate	31.3 × 26	LS, SEM	(Yang et al. 2001)

* P = Polar axis, E = Equatorial diameter

2.6 ECOLOGY

Rheum L., an extensively diversified and radiated genus, occurs mainly in mountainous regions of the QTP and its adjacent areas (Wan et al. 2011). Putatively, these regions form the centers of both origin and diversification of the genus because it shows highly diversified morphology as well as substantial endemism in these areas, both at species as well as section levels (Wan et al. 2011).

Rheum has undergone rapid radiations, probably due to immense uplifts of the QTP as depicted from the dispersion and the ancestral area reconstruction analyses (Wan et al. 2011; Sun et al. 2012). A remarkable phenotypic diversification is clearly visible in rhubarb as a measure of its adaptation to different habitat alterations (Wan et al. 2014). For instance, several species have acquired dwarf stature and possess either drooping bracts or coriaceous basal leaves. The evolutionary reward for decumbent forms is believed to be the protection against harsh winds, and the role of drooping bracts is thought to be in pollen protection or in temperature maintenance of inflorescence under low temperature stress as well as from ultraviolet radiation, hence making it possible for these forms to be dispersed along the snowline up to an altitudee of 5000 m (Xie 2000). In some species, degenerated leaves occur around the stem, and to minimize the transpiration rate, basal leaves are often covered with indumentum or verruca, in turn preventing the plant from withering so much so that the plant grows in the Gobi Desert (Liu et al. 2013). In addition, *R. palaestinum*, for instance, has evolved its leaves in way to collect rainwater, by being broad, rigid, waxy, and bearing channels cut into them and also with adequate power to cause deep soil penetration (Lev-Yadun et al. 2009).

The majority of *Rheum* species prefer habitats like the dry and cold meadows of alpine regions, dry slopes, and steppe deserts. Owing to the adaptability for arid habitats, similar morphologies have been evolved among the species from various lineages of the genus that are generally treated as morphologically diverse *Rheum* (Wan et al. 2011). Generally, well-drained, fertile, and organically rich soil is optimum for proper cultivation of rhubarb, maintaining a pH of 6.0–6.8 (Roggemans and Boxus 1988).

Various phenological episodes in the lifecycle of the plant provide the information of the altitudinal gradient and also about the pollination requirements, helping one to know about the behavior and mutualistic dealings of the plant, hence aiding in devising the precise planning for proper cultivation and conservation practices (Wani et al. 2009). Among different environmental cues, temperature was found to have prominent effect on different phenophases of the plant. Throughout the subzero temperatures of winter months, the rhizomes remain dormant under soil (November to March/April). More often, the plant grows among *Juniperus squamata* naturally and outcompetes it in height by exposing its spike/inflorescence with the help of nodes and internodes (Wani et al. 2009).

Yet another possible reason believed to drive the high diversification of the temperate genera in the regions of the QTP may have been the convergent evolution under habitat pressure, though the systematic position of many sections is still ambiguous; for instance, Sect. Globulosa possessing globular inflorescence, and Sect. Nobilia having semi-translucent bracts. Moreover, regarding the classification of the genus on the basis of exine patterns, substantial controversies have come into place, due to recent advances in palynological research (Wang et al. 2005).

The altitudinal ranges and the threat status of different species of the genus vary considerably. For instance, the two species, viz. *R. palmatum* and *R. tanguticum* overlap in their habitats in north-western China (Wang et al. 2010) and usually occur along the rivers in the valleys and also in the foothills in the forest edges (Wang et al. 2012). Similarly, an altitude of 1100 to 4600 m is the possible ecological amplitude for *R. officinale* and on any mountain its altitudinal range does not exceed 600 m and may usually be found near the forest edges. As the habitat deterioration of *R. officinale* is alarming, the number of individuals of this species is rapidly diminishing annually, thereby destroying its resources in the wild. Hence, it can be anticipated that *R. officinale* is going to be another original medicinally important rhubarb plant that will be endangered (Wang et al. 2012). Another species, viz. *R. australe* growing at an altitudinal range of 2800–3000 m, is endemic to the Himalayan region dispersed widely in the temperate and subtropical regions from Kashmir to Sikkim (Pandith et al. 2014, 2018). The ecological niches of the species are found in the alpine region on rocky soils, crevices, and near streams and between boulders (Radhika et al. 2010). The species is endangered in the Himalayan regions of the Kashmir Valley due to its extensive exploitation and needs to be effectively conserved to prevent its extinction (Kabir Dar et al. 2015; Pandith et al. 2018).

REFERENCES

Agarwal S, Singh SS, Verma S, Kumar S (2000) Antifungal activity of anthraquinone derivatives from Rheum emodi. *Journal of Ethnopharmacology* 72(1–2):43–46.

Airy Shaw H (1973) *A Dictionary of the Flowering Plants and Ferns.* Cambridge: CUP.

Ali SI (1980) *Flora of Pakistan.* Islamabad, Pakistan: Pakistan Agricultural Research Council.

Ayodele A, Olowokudejo J (2006) The family Polygonaceae in West Africa: Taxonomic significance of leaf epidermal characters. *South African Journal of Botany* 72(3):442–459.

Baas P (2017) Mabberley's Plant-book–a portable dictionary of plants, their classification and uses. DJ Mabberley. 1102 pp., 2017. Cambridge University Press. Price: EUR 69.00 or GBP 59.99 (hardback). ISBN 978-1-107-11502-6. *IAWA Journal* 38(4):573–573.

Bentham G, Hooker J (1880) Polygonaceae. *Genera Plantarum* 3:88–105.

Bessey CE (1915) The phylogenetic taxonomy of flowering plants. *Annals of the Missouri Botanical Garden* 2(1/2):109–164.

Brandbyge J (1993) Polygonaceae. In: *Flowering Plants Dicotyledons.* Berlin/Heidelberg: Springer:531–544.

Byng JW, Chase MW, Christenhusz MJ, Fay MF, Judd WS, Mabberley DJ, Sennikov AN, Soltis DE, Soltis PS, Stevens PF (2016) An update of the Angiosperm Phylogeny Group classification for the orders and families of flowering plants: APG IV. *Botanical Journal of the Linnean Society* 181(1):1–20.

Botany and Ecology

Christenhusz MJ, Byng JW (2016) The number of known plants species in the world and its annual increase. *Phytotaxa* 261(3):201–217.

Dammer U (1887) *Polygonaceae*. W. Engelmann. The Natural Plant Families, Leipzig, Germany.

de Jussieu A-L (1789) *Genera plantarum secundum ordines naturales Disposita juxta methodum in horto regio parisiensi exaratam, anno 1774*. Paris: veuve Herissant.

Erdtman G (1952) *Pollen Morphology and Plant Taxonomy: An Introduction to Palynology. Angiosperms*, Volume 1. New York: Hafner.

Erdtman G (1969) *Handbook of Palynolgy: Morphology, Taxonomy, Ecology. An Introduction to the Study of Pollen Grains and Spores*. New York: Hafner.

Ferguson IK (1985) *The Role of Pollen Morphology in Plant Systematic*. Kew, Richmond, UK.

Freeman CC, Reveal J (2005) *Flora of North America*, Volume 5. St Louis: Missouri Botanical Garden:492–496.

Gohil R, Rather G (1986) Cytogenetic studies of some members of Polygonaceae of Kashmir. III Rheum L. *Cytologia* 51(4):693–700.

Gross H (1912a) Arbeit aus d. Botan. Inst. d. Kgl. Albertus-Univ. zu Königsberg iP: Beiträge zur Kenntnis der Polygonaceen. Ph.D. diss., W. Engelmann.

Gross H (1912b) Beitr? ge zur Kenntnis der Polygonaceen. *Bot JB* 49:234–339.

Gupta RK, Bajracharya GB, Jha RN (2014) Antibacterial activity, cytotoxicity, antioxidant capacity and phytochemicals of Rheum australe rhizomes of Nepal. *Journal of Pharmacognosy and Phytochemistry* 2(6):125–128.

Hamayun M, Khan MA, Chudhary MF, Ahmad H (2007) Studies on traditional knowledge of medicinal herbs of Swat Kohistan, District Swat, Pakistan. *Journal of Herbs, Spices and Medicinal Plants* 12(4):11–28.

Haraldson K (1978) Anatomy and taxonomy in Polygonaceae subfam. Polygonoideae Meisn. emend. Jaretzky. *Acta Universitatis Upsaliensis*. 1:1–95

Hedberg O (1945) Pollen morphology in the genus Polygonum ls lat. and its taxonomic significance. *Sven Bot Tisdskr* 40:371–404.

Hooker JD (1897) *The flora of British India*, Volume 7. L. Reeve. London.

Huang Q, Lu G, Shen HM, Chung MC, Ong CN (2007) Anti-cancer properties of anthraquinones from rhubarb. *Medicinal Research Reviews* 27(5):609–630.

Huxley AJ, Griffiths M (1999) *New Royal Horticultural Society Dictionary of Gardening*. Macmillan, London: Grove's Dictionaries Inc.

Jaretzky R (1925) Beiträge zur Systematik der Polygonaceae unter Berücksichtigung des Oxymethylanthrachinon-Vorkommens. *Repertorium Novarum Specierum Regni Vegetabilis* 22(4–12):49–83.

Jin W, Wang YF, Ge RL, Shi HM, Jia CQ, Tu PF (2007) Simultaneous analysis of multiple bioactive constituents in Rheum tanguticum Maxim. ex Balf. by high-performance liquid chromatography coupled to tandem mass spectrometry. *Rapid Communications in Mass Spectrometry: An International Journal Devoted to the Rapid Dissemination of Up-to-the-Minute Research in Mass Spectrometry* 21(14):2351–2360.

Judd W, Campbell CS, Kellogg E, Stevens P (1999) *Plant Systematics. A Phylogenetic Approach*. Sunderland, MA: Sinauer Associates:3–4.

Kabir Dar A, Siddiqui M, Wahid-ul H, Lone A, Manzoor N, Haji A (2015) Threat status of rheum emodi-A study in selected cis-Himalayan regions of Kashmir valley Jammu & Kashmir India. *Medicinal and Aromatic Plants* 4:183.

Kao T, Cheng C-Y (1975) Synopsis of the Chinese rheum. *Acta Phytotaxonom Sinica* 13(3):69–82.

Karthikeyan S (2000) *A Statistical Analysis of Flowering Plants of India. Flora of India-Introductory Volume.* Calcutta, India: Botanical Society of India, 201–217.

Kubo I, Murai Y, Soediro I, Soetarno S, Sastrodihardjo S (1992) Cytotoxic anthraquinones from Rheum pulmatum. *Phytochemistry* 31(3):1063–1065.

Lev-Yadun S, Katzir G, Neeman G (2009) Rheum palaestinum (desert rhubarb), a self-irrigating desert plant. *Naturwissenschaften* 96(3):393–397.

Li A (1998) *Flora reipublicae popularis Sinicae: Tomus 25 (1). Angiospermae, Dicotyledoneae, Polygonaceae.* Beijing: Science Press:237.-illus.. ISBN 703006450X Ch Icones, Keys. Geog.

Li J, Zhang J (1983) Investigation on origin and quality of commodities of rhubarb. *Chinese Journal of Pharmaceutical Analysis* 3(6):333–339.

Libert B, Englund R (1989) Present distribution and ecology of Rheum rhaponticum (Polygonaceae). *Willdenowia* 1:91–98.

Liu BB, Opgenoorth L, Miehe G, Zhang DY, Wan DS, Zhao CM, Jia DR, Liu JQ (2013) Molecular bases for parallel evolution of translucent bracts in an alpine "glasshouse" plant Rheum alexandrae (Polygonaceae). *Journal of Systematics and Evolution* 51(2):134–141.

Losina-Losinskaya A (1936a) The Genus Rheum and Its Species. In: *Acta Instituti Botanici Academiae Scientiarum URSS, Unionis Rerum, Publicarum Soveticarum Series, Socialisticarum Series 1, Fasciculus 3:67-141.* Moscow, Russia..

Lozina-Lozinskaja A (1936b) Sistematiceskij obzor dikorastuscich vidov roda Rheum L. In: *Trudy Botanicheskogo Instituta Akademii Nauk SSSR. Series 1.* Moscow, Russia: 67–141.

Medynska E, Smolarz HD (2005) Comparative study of phenolic acids from underground parts of Rheum palmatum L., R. Rhaponticum L. and R. undulatum L. *Acta Societatis Botanicorum Poloniae* 74(4):275–279.

Miraj S (2016) Therapeutic effects of Rheum palmatum L.(Dahuang): A systematic review. *Der Pharma Chemica* 8(13):50–54.

Mondal M (1997) *Pollen Morphology and Systematic Relationship of the Family Polygonaceae*, Volume 1. Calcutta, India: Botanical Survey of India: 86–94.

Muller J (1981) Fossil pollen records of extant angiosperms. *The Botanical Review* 47(1):1.

Munshi AH, Javeid G (1986) *Systematic Studies in Polygonaceae of Kashmir Himalaya.* J Econ Taxon Bot add ser 2. Jodhpur, Rajasthan, India: Scientific Publishers.

Nowicke JW, Skvarla JJ (1977) Pollen morphology and the relationship of the Plumbaginaceae, Polygonaceae, and Primulaceae to the order Centrospermae. *Smithsonian Contributions to Botany* 37:1–7.

Pandith SA, Dar RA, Lattoo SK, Shah MA, Reshi ZA (2018) Rheum australe, an endangered high-value medicinal herb of North Western Himalayas: A review of its botany, ethnomedical uses, phytochemistry and pharmacology. *Phytochemistry Reviews: Proceedings of the Phytochemical Society of Europe* 17(3):573–609.

Pandith SA, Hussain A, Bhat WW, Dhar N, Qazi AK, Rana S, Razdan S, Wani TA, Shah MA, Bedi Y, Hamid A, Lattoo SK (2014) Evaluation of anthraquinones from Himalayan rhubarb (Rheum emodi Wall. ex Meissn.) as antiproliferative agents. *South African Journal of Botany* 95:1–8.

Perdrigeat C-A (1900) Anatomie comparée des Polygonées et ses rapports avec la morphologie et la classification. *Actes de la Société Linnéenne de Bordeaux* 55:1–91.

Radhika R, Krishnakumar I, Sudarsanam D (2010) Antidiabetic activity of Rheum emodi in alloxan induced diabetic rats. *International Journal of Pharmaceutical Sciences and Research* 8:296–300.

Roggemans J, Boxus P (1988) Rhubarb (Rheum rhaponticum L.). *Crops II* Bajaj YPS (editor). Berlin Heidelberg, Berlin, Heidelberg: Springer:339–350. doi:10.1007/978-3-642-73520-2_16

Botany and Ecology

Samappito S, Page JE, Schmidt J, De-Eknamkul W, Kutchan TM (2003) Aromatic and pyrone polyketides synthesized by a stilbene synthase from Rheum tataricum. *Phytochemistry* 62(3):313–323.

Sanchez A (2011) *Evolutionary Relationships in Polygonaceae with Emphasis on Triplaris.* Wake Forest University. ProQuest Dissertations Publishing, North Carolina.

Sanchez A, Kron KA (2008) Phylogenetics of Polygonaceae with an emphasis on the evolution of Eriogonoideae. *Systematic Botany* 33(1):87–96.

Sanchez A, Kron KA (2009) Phylogenetic relationships of Afrobrunnichia Hutch. & Dalziel (Polygonaceae) based on three chloroplast genes and ITS. *Taxon* 58(3):781–792.

Sanchez A, Schuster TM, Kron KA (2009) A large-scale phylogeny of Polygonaceae based on molecular data. *International Journal of Plant Sciences* 170(8):1044–1055.

Santapau H, Henry AN (1973) *A Dictionary of the Flowering Plants in India*, Volume vii. New Delhi: CSIR:198. Geog 6.

Srivastava R (2014) Family Polygonaceae in India. *Indian Journal of Plant Sciences* 3(2):112–150.

Stewart RR, Ali S, Nasir E (1972) *An Annotated Catalogue of the Vascular Plants of West Pakistan and Kashmir.* printed at Karachi, Pakistan: Fakhri Print Press.

Sun Y, Wang A, Wan D, Wang Q, Liu J (2012) Rapid radiation of Rheum (Polygonaceae) and parallel evolution of morphological traits. *Molecular Phylogenetics and Evolution* 63(1):150–158.

Uddin K, Rahman A, Islam A (2014) Taxonomy and traditional medicine practices of Polygonaceae (smartweed) family at Rajshahi, Bangladesh. *International Journal of Advanced Research* 2(11):459–469.

Van Leeuwen P, Punt W, Hoen P (1988) The northwest European pollen flora, 43: Polygonaceae. *Review of Palaeobotany and Palynology* 57(1–2):81.

Wake DB, Wake MH, Specht CD (2011) Homoplasy: From detecting pattern to determining process and mechanism of evolution. *Science* 331(6020):1032–1035.

Walker JW (1974) Evolution of exine structure in the pollen of primitive angiosperms. *American Journal of Botany* 61(8):891–902.

Wan D, Sun Y, Zhang X, Bai X, Wang J, Wang A, Milne R (2014) Multiple ITS copies reveal extensive hybridization within Rheum (Polygonaceae), a genus that has undergone rapid radiation. *PLOS ONE* 9(2):e89769.

Wan D, Wang A, Zhang X, Wang Z, Li Z (2011) Gene duplication and adaptive evolution of the CHS-like genes within the genus Rheum (Polygonaceae). *Biochemical Systematics and Ecology* 39(4–6):651–659.

Wang A, Yang M, Liu J (2005) Molecular phylogeny, recent radiation and evolution of gross morphology of the rhubarb genus Rheum (Polygonaceae) inferred from chloroplast DNA trn LF sequences. *Annals of Botany* 96(3):489–498.

Wang F, Qian N, Zhang Y, Yang H (1995) *Pollen Morphology in China.* Beijing: Science Press.

Wang X-M, Hou X-Q, Zhang Y-Q, Li Y (2010) Distribution pattern of genuine species of rhubarb as traditional Chinese medicine. *Journal of Medicinal Plants Research* 4(18):1865–1876.

Wang X, Ren Y (2009) Rheum tanguticum, an endangered medicinal plant endemic to China. *Journal of Medicinal Plants Research* 3(13):1195–1203.

Wang X, Yang R, Feng S, Hou X, Zhang Y, Li Y, Ren Y (2012) Genetic variation in Rheum palmatum and Rheum tanguticum (Polygonaceae), two medicinally and endemic species in China using ISSR markers. *PLOS ONE* 7(12):e51667.

Wani PA, Nawchoo IA, Wafai B (2009) The role of phenotypic plasticity, phenology, breeding behaviour and pollination systems in conservation of Rheum emodi Wall. exMeisn.(Polygonaceae)—A Threatened Medicinal Herb of North West Himalaya. *The International Journal of Plant Reproductive Biology* 1(2):179–189.

Wodehouse RP (1931) Pollen grains in the identification and classification of plants VI. Polygonaceae. *American Journal of Botany* 18(9):749–764.

Wodehouse RP (1935) *Pollen Grains.* New York; London: Mcgraw-Hill Book Company, Inc.

Wu Z, Raven PH, Hong D (2003) Flora of China. *Ulmaceae through Basellaceae*, Volume 5. Beijing, China: Science Press.

Xiao P, He L, Wang L (1984) Ethnopharmacologic study of Chinese rhubarb. *Journal of Ethnopharmacology* 10(3):275–293.

Xie Z (2000) *Eco-Geographical Distribution of the Species from Rheum L. Polygonaceae in China. China's Biodiversity Conservation toward the 21st Century.* Beijing, China: China Forest Press.

Yang M, Zhang D, Zheng J, Liu J (2001) Pollen morphology and its systematic and ecological significance in Rheum (Polygonaceae) from China. *Nordic Journal of Botany* 21(4):411–418.

Zhang X, Zhou Z (1998) *A Study on Pollen Morphology and Its Phylogeny Of 'Polygonaceae' in China.* Beijing, China: China Science and Technology University Press.

3 Traditional Uses

The human practice of using plants or plant parts as a source of traditional medicine is primarily grounded on information and familiarity that has been handed down through generations. The prodigious social and technological changes in the 15th century led to the compilation and documentation of knowledge of medicinal herbs which indeed became the prototypes of various pharmacopoeias and many other books on herbal medicine (Adams et al. 2009). The majority of the population (70–80% of the human population, tribals in particular), in developing countries is dependent on traditional herbal medicine for primary healthcare. In fact, varied plants with medical efficacy have a long history of frequent use in different traditional systems of medicine against various ailments. Moreover, many people in the developed world also seem to have an inclination toward medicinal herbs as they believe "natural is better," and there are fewer reports of adverse side effects with these natural remedies (Lewis 2003). But, with respect to their use across traditions, few written documents are available and these are difficult to access. Supplementing this loophole, contemporary researchers generally show less attention to exploring this rich and natural scientific wealth at the required and anticipated level. Nevertheless, there are promising opportunities in this field wherein recent attention paid by traditional Chinese medicine serves as an eye-opener. Moreover, in the recent past, as an attempt to evaluate the utility of traditional medicinal knowledge even well-known pharmaceutical companies like Novartis have invested about US$100 million in establishing a novel research and development center in Shanghai that is devoted to set up a common platform for age-old medicinal concepts and the approaches of current biomedical research (Stone and Xin 2006). The same Novartis company, a Swiss drug giant, in collaboration with the World Health Organization (WHO), has produced and even distributed the most advanced anti-malarial drug, Coartem® (Riamet®), derived from a sesquiterpene peroxide artemisinin, originally isolated in 1971 by Chinese scientists from *Artemisia annua* (White 2008). Additionally, it is also reported that about 62 million Coartem® treatment courses, which were distributed in almost 30 countries, saved a population of nearly 200,000 in 2006 (Adams et al. 2009). During the ten-year time period for the memorandum of understanding (MoU) between Novartis Pharma AG and the WHO for the supply of Coartem® (artemether/lumefantrine, AL), about 400 million treatment courses of Coartem® were provided to malaria-endemic developing countries mainly for children, with a price which was fairly reduced by 43–60%. In fact, and following international licensing approval in 1999, Coartem® became the platform to treat malaria at the global level with a registered use in nearly 90 countries. Furthermore, the unanticipated market demand for AL is also known to have compelled the scientific community to a melodramatic advance toward quality production of *A.*

DOI: 10.1201/9780429340390-3

annua in plant farming in Kenya and China at the commercial scale in addition to instituting widespread and new production facilities in Africa, China, and the United States (Premji 2009). Therefore, ethnomedicine seems to play an important role in current drug discovery programs and is becoming a potential source for a good number of modern drugs developed against old and emerging diseases. In fact, the generation of new and potential lead molecules as possible drug agents seems plausible if the contemporary scientific community becomes motivated to uncover forgotten historic medical works and their associated remedies.

Rhubarb (Polygonaceae), with an average life span of five to eight years, has a long and often oral history of herbal utility in Asian traditional medical systems, as well as in Europe, and surprisingly, even in America. The prime outcome of rhubarb as a herbal medicine is actually a constructive and complementary impact on the digestive system. A property of the rhubarb root makes it a very effective laxative. Its astringent quality improves the bowel tone once it purges the intestines, making it an effective agent for the digestive tract. Indeed, the laxative effects of rhubarb have made it a valued aid to treat some common health issues like gastroenteritis, diarrhea, heartburn, stomach pain, constipation, and hemorrhoids, etc. Certain people also make use of rhubarb to minimize strain during bowel movements which thereby reduces the related pain from hemorrhoids or anal fissures. Sometimes, rhubarb is also applied to the skin to treat winter-associated cold sores. Pertinently, The Herbal Medicinal Product Committee (HMPC)—with exclusive authority at European level for the independent assessment of the registration of herbal medicines—has recognized the administration of rhubarb roots to treat sporadic constipation as a "well-established medicinal use." Moreover, the external use of alcoholic rhubarb root extracts is also known to be used for brushing for gum inflammation and against the oral mucosa (Moore 1997; Maclean and Taylor 2000; Gardener and Ghronicle 2014; Clementi and Misiti 2010).

Rhubarb is one of the most widely used medicinal herbs in traditional Chinese medicine. Its varied uses were mainly documented in the Divine Husbandman's classic of materia medica that was compiled in around AD 200 during the Han Dynasty. Da huang (Chinese rhubarb) is regarded as the most ancient and best-known plants used in Chinese herbal medicine. Da huang is actually formulated from the rhizome/roots of different species of rhubarb (*Rheum*) which include *R. tanguticum* Maxim. ex Balf., *R. palmatum* L., *R. officinale* Baill., or *R. undulatum*. It finds initial documentation primarily in the oldest Chinese materia medica, *Shen Nong's Herbal Classic* (*Shen Nong Ben Cao Jing*) which followed *Huangdi's Internal Classic* (*Huang Di Nei Jing*) with implications in the *Treatise on Febrile Diseases* (*Shan Han Lun*) as a laxative and bactericidal agent to rinse out the body, improve blood circulation, and to ease fever (Lai et al. 2015). Mature plants (at least six years old) are harvested in the fall season for the roots/rhizome which are later dried for use in various forms and against a series of ailments ranging from simple laxatives to antiseptic, aperient, diuretic, demulcent, purgative, stomachic, and as antispasmodic agents. Although it has the same geographical source, rhubarb has various trade names in commerce, viz. Chinese, East Indian/Himalayan, Russian, and Turkish. Importantly, the commercial/trade names of

Traditional Uses

rhubarb, the wonder drug, merely indicate the route through which it had reached the markets of Europe. Hayward, an Oxfordshire pharmacist in England in the 1760s, initiated the development and growth the rhubarb usually grown today. Indeed, it was only in the 1820s when rhubarb first landed in America, entering the country through Maine and Massachusetts to move west with the immigrants. European herbalists had suggested a promising use of rhubarb as a laxative and diuretic agent, besides having other potentialities to treat gout, kidney stones, and liver ailments, viz. jaundice, etc. Small doses are suggested for use against dysenteric diarrhea while the larger doses are known to show a laxative effect by causing purging that removes toxins from the intestinal tract, normally within eight hours after administration. Studies of the current investigations on rhubarb in growing economies like Japan and China have confirmed that it has the ability to suspend or halt the advancement of prolonged kidney failure. Besides having proven effects in endometriosis and certain menstrual issues, lindleyin, an important metabolite from the rhubarb rhizome, is known to display anti-inflammatory action with a fever-diminishing property analogous to that of aspirin. Herbal specialists in China make use of rhubarb root to treat ailments and disorders in the upper body which include, to mention a few, sinus with lung infections, eye contagions, and hemorrhages, etc. Chinese herbalists also use rhubarb root for diseases and disorders in the upper body, including sinus and lung infections, nosebleeds, and eye infections (Ashnagar et al. 2007). People using this drug also claim rhubarb to boost their appetite when administered in small amounts before taking meals. Further, it is also projected to endorse blood circulation, relieve pain in inflammation/injury, prevent bowel infections, treat burns, and can even minimize autoimmune reactions (Grieve and Leyel 1994; Hoffmann 1996; Mantani et al. 2002; Oi et al. 2002; Castleman 1991). Indeed, some studies promote rhubarb as an effective agent for sepsis in human and animal models (Lai et al. 2015). Nevertheless, all these claims need to be further ascertained while following proper and standard operating procedures.

As Chinese rhubarb has found a good domestic market and in some other countries around the world, Turkish rhubarb similarly found a wide and effective use as a potent medicinal herb in some of the earlier civilizations. Importantly, this perennial herb has a wide use in contemporary conventional and herbal medical systems (Chevallier 2000). The preliminary accounts of this traditional medicinal plant are documented in prehistoric Chinese writings which date back to 2700 BC (Chmelik 1999). While going through the pages of Chinese history, it seems that Turkish rhubarb has an even older history than that of Chinese rhubarb with its uses documented against fever as well as the purging effects. The herb has several historical episodes associated with it: It was taken as a remedy for fever by the emperor of the Liang dynasty (557–579), as a gift-bearing means, again to reduce the feverish conditions for the long-ruling Tang dynasty's emperor (618–907), as a source to fight an epidemic during the reign of the Song dynasty (960–1127), and used by a general from the Ming dynasty (1368–1644) as a suicide measure (Foster and Johnson 2008). Several effects of this rhubarb have been noted so far, viz. laxative, anti-viral, anti-inflammatory, and anti-oxidative, as well as its effectiveness against various other human and animal ailments which make it an exceptionally treasured medicinal source and a potent phytoceutical (Zhou et al. 2015; Liu et al. 2015; Schörkhuber et al. 1998). Moreover, and in view of the number of studies available pertaining to the wide range of benefits without definite adverse side effects, this herbal drug seems, and would be, an instrumental supplementation to any eco- and human-friendly herbal program. Interestingly, Indian/Himalayan rhubarb (*R. australe*) is one of the most sought-after species from the genus *Rheum* with wide pharmacological significance, as reported in the literature available to date. More or less, the ethnopharmacology of some of the reputed medicinal herbs from this genus of 60 extant species is similar. In that context, and for a detailed report on the ethnobotany of *Rheum* in general, a couple of comprehensive reviews on one of the promising species, *R. australe*, can be referred to: One review available is by Rokaya et. al. (2012) and the other was recently compiled by us (Pandith et al. 2018).

3.1 RHUBARB AS A FOOD PLANT

Rhubarb is an unfamiliar vegetable, officially classified as fruit by the United States Department of Agriculture (USDA) (https://fdc.nal.usda.gov/). Though a common garden plant in Northern Europe and North America, rhubarb is chiefly found in the temperate mountainous regions around the world; north-east Asia in particular. Culinary or garden rhubarb (*Rheum x hybridum* in the Royal Horticultural Society's list of recognized plant names) is the most common variety in the Western world. Owing to the availability of suitable greenhouse production, rhubarb is nowadays produced in many areas and is available nearly all year. The plant is cultivated for its fleshy petioles (rhubarb sticks or stalks) which are ready for consumption immediately after harvesting. There is a color range of rhubarb stalks from pale green to pink to red (more prevalent for consumption) with evenness analogous to that of celery, though it has no association with its appropriateness for cooking (Clementi and Misiti 2010). Rhubarb is possibly among the most sour-tasting vegetables in use. In fact, the herb is well often known for its thick stalks and sour taste which hinders its use in raw form while encouraging the use of some sweetener. The acidity of this actually medicinal plant is primarily due to the advanced levels of malic and oxalic acid (Rumpunen and Henriksen 1999). Fascinatingly, if grown in darkness, rhubarb becomes more tender with a lower level of sourness. The variety developed (in spring or late winter) in such a way is recognized as "forced rhubarb." The sour taste of this vegetable herb is usually avoided by cooking it with sugar. Indeed, the ready availability of sugar due to its cheapness in the 18th century made rhubarb a common food in households. Moreover, and to avoid the bitter leaves, rhubarb stalks have varied uses, and are preferably consumed (while adding plenty of sugar, in general) in the form of jams, pies, sauces, crumbles, tarts, sweet soups, cocktails, and as rhubarb wine or an ingredient in baked goods. When stewed, the stalks yield a tart sauce which can be consumed with sugar. However, rhubarb recipes with little or no added sugar are also available. In North America and the United

Kingdom, a traditional dessert in the form of sweet rhubarb pies is usually consumed giving the plant a new name as the "pie plant."

Rhubarb, although low in calories, is known to be a good source of some of the vital minerals (K, Ca) and vitamins (K, C) as well, in addition to the carbohydrates and proteins with similar amounts of dietary fiber as found in apples, oranges, or celery. Further, being a rich source of vitamin K1, it is supposed to provide nearly 26–37% of an individual's "daily value" (DV) in a 100-gram serving depending on the mode of administration (preferably when cooked). As per the USDA Nutrient Database, rhubarb has 0% cholesterol content (Pandith et al. 2018) (https://fdc.nal.usda.gov/). While supplemented with sugar, a 100-gram serving of cooked rhubarb has been found to contain: 116 calories, 31.2 g of carbs, 2 g of fiber, 0.4 g of proteins, 26% DV of vitamin K1, 6% DV of vitamin C, 15% DV of calcium, 3% DV of potassium, and 1% DV of folate (https://fdc.nal.usda .gov/). Importantly, rhubarb seems to be a good source of calcium; but unfortunately, the mineral/nutrient is in a non-dietary/anti-nutrient form of calcium oxalate (CaOx), the form which the human body is unable to absorb proficiently (Heaney and Weaver 1989). Indeed, rhubarb is known to be one of the richest dietary sources of CaOx, the very common form of oxalic acid in different plant species with an established role in calcium regulation. The oxalic acid concentration is supposed to augment from spring to the summer season due to which folk tradition suggests that its harvesting should be done before the month of June (average flowering period of the plant). The CaOx levels may vary between different species of *Rheum* and between different parts (leaves, petioles) of the same rhubarb species. Higher levels of CaOx can cause hyperoxaluria, a severe condition characterized by the accretion of CaOx crystals in different organs of the body. These CaOx crystals sometimes might form kidney stones which may, in the long run, even lead to kidney failure (Albersmeyer et al. 2012). It is therefore suggested, by herbalists and dieticians, that rhubarb (preferably stalks while avoiding the leaves with high CaOx content) may be consumed in a balanced way and in cooked form as cooking diminishes its CaOx content by 30–87%. Additionally, although rhubarb is a good source of antioxidants, vitamins, and fiber, it would be better avoided by people who are disposed to kidney stones (Chai and Liebman 2005; Sanz and Reig 1992; Tallqvist and Vaananen 1960).

In conclusion, rhubarb, as discussed above, is a high-value medicinal herb with a great therapeutic reputation, besides being used as a food/vegetable source when in season. Indeed, different species of *Rheum* find good household use as a common vegetable/food source (especially in winter among tribals) in the regions where it grows. There is also positive evidence of its use as a potential source of medicine since ancient times to cure a wide range of health issues that are devoid of any recognized adverse effects. Certainly, the literature available with regard to this wondrous drug justifies its significant medical implications when used alone or as an adjunct in different herbal formulations as stated in various traditional systems of medicine, including Ayurveda, Tibetan, Homeopathic, Chinese, and Unani medical systems. It is pertinent to mention that the medical efficacy of *Rheum* against different health ailments (particularly in humans) is due to the presence of some of the major phytoconstituents, viz. anthraquinones and stilbenoids, etc., with proven pharmacological effectiveness toward a range of human disorders/diseases. A series of novel and potent derivatives of these bioactive phytochemical constituents has been generated and later assessed for better anti-malarial, anti-microbial, anti-oxidative, and anti-proliferative activities while utilizing contemporary systems and synthetic biology approaches. Nevertheless, more and progressive investigations are necessary to further advance this valuable, wondrous drug which may include extraction and isolation of bioactive compounds including the production of highly effective novel derivatives to screen and authenticate the traditional knowledge of rhubarb. Similarly, the real mechanism (*in vitro* and *in vivo*) of action of the different herbal formulations using rhubarb (individually or as a supplement) in dried/powder or extract form needs to be screened and understood. Additionally, serious examination of the toxicity levels, pharmacokinetic and pharmacodynamic mechanisms, and bioavailability are the issues which need attention to modulate the crucial bioactive constituents and/or their potent derivatives from *Rheum* as fundamental frameworks for forthcoming drugs. Furthermore, due to the wide and well-known pharmacological efficacy of the low molecular weight secondary metabolites (as bioactives) so far isolated from different species of *Rheum*, their improved production in homo- and/or heterologous host systems has a vibrant significance in the field of secondary metabolite pathway engineering as attempted in various related studies across the world, including some of our own efforts in our lab on different medicinal plant species (Dhar et al. 2015; Pandith et al. 2016).

REFERENCES

Adams M, Berset C, Kessler M, Hamburger M (2009) Medicinal herbs for the treatment of rheumatic disorders—A survey of European herbals from the 16th and 17th century. *Journal of Ethnopharmacology* 121(3):343–359.

Albersmeyer M, Hilge R, Schröttle A, Weiss M, Sitter T, Vielhauer V (2012) Acute kidney injury after ingestion of rhubarb: Secondary oxalate nephropathy in a patient with type 1 diabetes. *BMC Nephrology* 13(1):141.

Ashnagar A, Naseri NG, Nasab HH (2007) Isolation and identification of anthralin from the roots of rhubarb plant (Rheum palmatum). *E-Journal of Chemistry* 4(4):546–549.

Castleman M (1991) The healing herbs: The ultimate guide to the curative power of nature's medicine. *Pensylvania: G and C Merriam* 1:128–135.

Chai W, Liebman M (2005) Effect of different cooking methods on vegetable oxalate content. *Journal of Agricultural and Food Chemistry* 53(8):3027–3030.

Chevallier A (2000) *Natural Health Encyclopedia of Herbal Medicine.* London, England: Dorling Kindersley Limited (DK).

Chmelik S (1999) *Chinese Herbal Secrets: The Key to Total Health.* London: Penguin.

Clementi EM, Misiti F (2010) Potential health benefits of rhubarb. In: *Bioactive Foods in Promoting Health.* Amsterdam, The Netherlands: Elsevier:407–423.

Dhar N, Razdan S, Rana S, Bhat WW, Vishwakarma R, Lattoo SK (2015) A decade of molecular understanding of withanolide biosynthesis and in vitro studies in Withania somnifera (L.) Dunal: Prospects and perspectives for pathway engineering. *Frontiers in Plant Science* 6:1031.

Traditional Uses 45

Foster S, Johnson RL (2008) *National Geographic. Desk Reference to Nature's Medicine.* Washington, DC: National Geographic Society.

Gardener C, Ghronicle G (2014) *Gardener's magazine. Rhubarb: The wondrous drug,* Volume 191. New Jersey: Princeton:321.

Grieve M, Leyel C (1994) *The Medicinal, Culinary, Cosmetic and Economic Properties, Cultivation and Folklore of Herbs, Grasses, Fungi, Shrubs and Trees with All Their Modern Scientific Uses. A Modern Herbal Tiger Books International.* Chatham, Kent: Mackays of Chatham, PLC:961–962.

Heaney R, Weaver CM (1989) Oxalate: Effect on calcium absorbability. *The American Journal of Clinical Nutrition* 50(4):830–832.

Hoffmann D (1996) *The Complete Illustrated Herbal: A Safe and Practical Guide to Making and Using Herbal Remedies.* Barnes & Noble Books. Harper Collins Publishers, New York.

Lai F, Zhang Y, Xie D-P, Mai S-T, Weng Y-N, Du J-D, Wu G-P, Zheng J-X, Han Y (2015) A systematic review of rhubarb (a Traditional Chinese Medicine) used for the treatment of experimental sepsis. *Evidence-Based Complementary and Alternative Medicine* (doi: 10.1155/2015/131283).

Lewis WH (2003) Pharmaceutical discoveries based on ethnomedicinal plants: 1985 to 2000 and beyond. *Economic Botany* 57(1):126.

Liu Z, Ma N, Zhong Y, Yang Z-Q (2015) Antiviral effect of emodin from Rheum palmatum against coxsakievirus B 5 and human respiratory syncytial virus in vitro. *Journal of Huazhong University of Science and Technology [Medical Sciences]* 35(6):916–922.

Maclean W, Taylor K (2000) *Clinical Manual of Chinese Herbal Patent Medicines.* London: Pangolin Press.

Mantani N, Sekiya N, Sakai S, Kogure T, Shimada Y, Terasawa K (2002) Rhubarb use in patients treated with Kampo medicines-A risk for gastric cancer? *Yakugaku Zasshi* 122(6):403–405.

Moore M (1997) *Herbal Formulas for the Clinic and Home.* Bisbee, AZ: Southwest School of Botanical Medicine.

Oi H, Matsuura D, Miyake M, Ueno M, Takai I, Yamamoto T, Kubo M, Moss J, Noda M (2002) Identification in traditional herbal medications and confirmation by synthesis of factors that inhibit cholera toxin-induced fluid accumulation. *Proceedings of the National Academy of Sciences of the United States of America* 99(5):3042–3046.

Pandith SA, Dar RA, Lattoo SK, Shah MA, Reshi ZA (2018) Rheum australe, an endangered high-value medicinal herb of North Western Himalayas: A review of its botany, ethnomedical uses, phytochemistry and pharmacology. *Phytochemistry Reviews: Proceedings of the Phytochemical Society of Europe* 17(3):573–609.

Pandith SA, Dhar N, Rana S, Bhat WW, Kushwaha M, Gupta AP, Shah MA, Vishwakarma R, Lattoo SK (2016) Functional promiscuity of two divergent paralogs of type III plant polyketide synthases. *Plant Physiology* 171(4):2599–2619.

Premji ZG (2009) Coartem®: The journey to the clinic. *Malaria Journal* 8(1):S3.

Rokaya MB, Münzbergová Z, Timsina B, Bhattarai KR (2012) Rheum australe D. Don: A review of its botany, ethnobotany, phytochemistry and pharmacology. *Journal of Ethnopharmacology* 141(3):761–774.

Rumpunen K, Henriksen K (1999) Phytochemical and morphological characterization of seventy-one cultivars and selections of culinary rhubarb (Rheum spp.). *The Journal of Horticultural Science and Biotechnology* 74(1):13–18.

Sanz P, Reig R (1992) Clinical and pathological findings in fatal plant oxalosis. A review. *The American Journal of Forensic Medicine and Pathology* 13(4):342–345.

Schörkhuber M, Richter M, Dutter A, Sontag G, Marian B (1998) Effect of anthraquinone-laxatives on the proliferation and urokinase secretion of normal, premalignant and malignant colonic epithelial cells. *European Journal of Cancer* 34(7):1091–1098.

Stone R, Xin H (2006) *Novartis Invests $100 Million in Shanghai*. American Association for the Advancement of Science.

Tallqvist H, Vaananen I (1960) Death of a child from oxalic acid poisoning due to eating rhubarb leaves. In: *Annales Paediatriae Fenniae*. Helsinki, JAMA: 144–147.

White NJ (2008) Qinghaosu (artemisinin): The price of success. *Science* 320(5874):330–334.

Zhou Y-X, Xia W, Yue W, Peng C, Rahman K, Zhang H (2015) Rhein: A review of pharmacological activities. *Evidence-Based Complementary and Alternative Medicine* (https://doi.org/10.1155/2015/578107).

4 Phytochemistry

And as for all the "patronage" of all the clowns and boors
That squint their little narrow eyes at any freak of yours,
Do leave them to your prosier friends—such fellows
ought to die when Rhubarb is so very scarce and ipecac so high!
                              —Oliver Wendell Holmes, *Nux Postcoenatica*

These lines infer that in 18th- and 19th-century Europe and America, rhubarb was being used as a highly desirable cathartic therapy and tonic, and also as food, especially in 19th-century Britain and America. In fact, several editions of the *Encyclopedia Britannica* are witness to the rise and fall in popularity of this wondrous aperient drug (and sour fruit/vegetable) (Foust 2014).

Pertinently, plants are a potential source of variety of secondary chemical constituents with wide pharmacological efficacy, as reported in several herbal/traditional pharmacopeia. These low-molecular-weight compounds are produced as natural protection agents against herbivory and microbial attack. Indeed, treating diseases with plants or plant preparations has been a common practice in virtually all cultures for centuries; and, in order to learn about the biological activity of their chemical constituents, several plant species have been the subject of scientific investigation for many years. This has endorsed the prospective generation of lead molecules as potent and safe natural, alternative, and antimicrobial drug agents/molecules (Kosikowska et al. 2010).

Rhubarb has a long history of being used to treat different ailments for centuries, and across the regions where it grows. The remedying properties of this plant are ascribed to some of its major phytoconstituents, particularly anthraquinones, stilbenoids, and flavonoids. The pharmacological efficacy of these metabolites and/or their derivatives is being scientifically investigated as lead molecules for the treatment of various diseases and ailments (Pandith et al. 2018). Recent investigations on the chemical constitution of different species of rhubarb have revealed that a wide variety of the derivatives of free classes of compounds such as anthracene derivatives (responsible for the purgative effect), stilbene glycosides, naphthalene derivatives, catechin derivatives, and some tannin-related compounds such as galloyl esters of glucose occur in their leaves and rhizomes (Krafczyk et al. 2008). The availability, quantity, and quality of these chemical compounds, however, is not equal and comparable in different species of rhubarb, and within different samples of the same species vis-à-vis their eco-physiological attributes. In fact, the concentration and number of compounds available in a rhubarb sample has been found to vary with the type of habitat, harvesting time, altitude, and soil properties, etc. For instance, in a study carried out by Malik et al., it was found that five compounds (chrysophanol glycoside, emodin glycoside, chrysophanol, emodin, and physcion) were present in 9- to 12-month-old plant

DOI: 10.1201/9780429340390-4

47

roots, whereas the six-month-old rhubarb roots showed the significant presence of only two compounds, viz. chrysophanol glycoside and emodin glycoside (Malik et al. 2010). Further, in some of our own studies on *R. australe* (Pandith et al. 2014, 2016, 2018) we have observed promising differences of major flavonoid/anthraquinone constituents at spatial and temporal levels; the study (Pandith et al. 2016) was relatively extended to *in vitro* propagated plants (data unpublished) as well, which also displayed similar results. These studies are evidence that the chemical constitution of a plant in nature and at a mature stage is non-comparable to the chemical constitution of young and *in vitro* regenerated/cultivated plants. Rokaya and others also concluded that the rhizome and leaves of rhubarb that are harvested early contain fewer metabolites, due to which they lose their effectiveness in herbal medicines (Rokaya et al. 2012a).

Additionally, several other studies have found that the concentrations of active constituents like flavonoids vary with respect to the altitude at which these plants grow. Some studies show that the concentration of these compounds was higher in plants growing at higher altitudes than plants growing at lower altitudes in the Himalayan region, as well as in other parts of the world. However, this trend is not always followed and varies with respect to different species (Rokaya et al. 2012a). Moreover, there is evidence of preferential metabolite channeling between the growth and defense mechanisms of a plant based on the nature and level of risk and the value of the particular tissue involved. Such trade-offs concerning growth and defense involve the routing of metabolites between secondary metabolite production and the introduction of enzymatic anti-oxidative defense systems (Abrol et al. 2012). Therefore, to understand the intricacies of these metabolites and the conditional/particular shift which they encounter within a specific plant species under varied environmental attributes/conditions, it is imperative to analyze and evaluate the production of these low-molecular-weight chemical compounds under natural and laboratory-controlled conditions.

The genus *Rheum* is enriched with a wide variety of chemical compounds which have been isolated from it and their structures elucidated. These chemical constituents belong to different classes of chemical compounds, viz., anthraquinones, anthocyanins, stilbenes, flavonoids, dianthrones, anthraglycosides, sterols, polyphenols, essential oils, organic acids, and vitamins, etc. (Agarwal et al. 2001). The phytoconstituents reported to date from various species of rhubarb across the regions where it grows are listed in detail in Table 4.1.

4.1 PHENOLICS

Flavonoids are a large group of polyphenolic secondary chemical constituents exhibiting wide diversity in their chemistry and occurrence. They are known to show ubiquitous occurrence in plants. Chemically, flavonoids consist of two benzene rings (A and B) linked by a heterocyclic pyrane or pyrone ring (ring C) or a three-carbon bridge (to form chalcones). So far, more than 4000 flavonoids have been identified in different species of plants. Based on the modification (hydroxylation, glycosylation, methoxylation, or prenylation) these are further divided

TABLE 4.1
Phytochemical Constituents Reported from Genus *Rheum*

Phytoconstituent	Species	Detection Method	Reference
Anthraquinones			
Emodin	S-1 to S-45	GC, GC-MS, HPLC, CA, CDFP, DAD, MS, ESI-MS, HPLC-MS, MLCCC, MS, NMR, FAS, SDS	(Agarwal et al. 2001; Tayade et al. 2012; Rehman et al. 2014; Kemper and Research 1999; Wang et al. 2011)
Aloe-emodin	S-1 to S-45	GC, GC-MS, HPLC, CA, CDFP, DAD, MS, ESI-MS, HPLC-MS, MLCCC, MS, NMR, FAS, SDS	(Agarwal et al. 2001; Ragasa et al. 2017; Tayade et al. 2012; Rehman et al. 2014; Kemper and Research 1999; Wang et al. 2011; Lin et al. 2008)
Chrysophanic acid	S-1 to S-45	GC, GC-MS, HPLC, CA, CDFP, DAD, MS, ESI-MS, HPLC-MS, MLCCC, MS, NMR, FAS, SDS	(Agarwal et al. 2001; Kemper and Research 1999; Tayade et al. 2012; Ragasa et al. 2017; Wang et al. 2011)
Chrysophanol	S-1 to S-45	GC, GC-MS, HPLC, CA, CDFP, DAD, MS, ESI-MS, HPLC-MS, MLCCC, MS, NMR, FAS, SDS	(Agarwal et al. 2001; Kemper and Research 1999; Tayade et al. 2012; Ragasa et al. 2017; Wang et al. 2011)
Physcion	S-1 to S-45	GC, GC-MS, HPLC, CA, CDFP, DAD, MS, ESI-MS, HPLC-MS, MLCCC, MS, NMR, FAS, SDS	(Agarwal et al. 2001; Kemper and Research 1999; Tayade et al. 2012; Ragasa et al. 2017; Wang et al. 2011)
Rhein	S-1 to S-45	GC, GC-MS, HPLC, CA, CDFP, DAD, MS, ESI-MS, HPLC-MS, MLCCC, MS, NMR, FAS, SDS	(Agarwal et al. 2001; Kemper and Research 1999; Tayade et al. 2012; Ragasa et al. 2017; Wang et al. 2011)
Istizin	S-6, S-7, S-9	GC, GC-MS, HPLC	(Agarwal et al. 2001)
Rheinal	S-6	GC, GC-MS, HPLC	(Agarwal et al. 2001; Kemper and Research 1999; Tayade et al. 2012; Ragasa et al. 2017; Wang et al. 2011)

(Continued)

TABLE 4.1 (CONTINUED)
Phytochemical Constituents Reported from Genus *Rheum*

Phytoconstituent	Species	Detection Method	Reference
Stilbenes			
Rhaponticin or Rhapontin	S-2, S-37, S-35, S-29, S-38, S-42, S-7, S-3, S4, S-27, S-43	GC, GC-MS, NMR, HPLC-DAD, MLCCC, FAS, SDS	(Agarwal et al. 2001; Ragasa et al. 2017; Tayade et al. 2012; Krafczyk et al. 2008)
Isorhapontin	S-45	FAB-MS, NMR	(Matsuda et al. 2001)
isorhapontegenin	S-45	FAB-MS, NMR	(Matsuda et al. 2001)
Deoxyrhaponticin (3,5-dihydroxy-4'-methoxystilbene 3-P-D-glucopyranoside)	S-5, S-38, S-42, S-36, S-27, S-43, S-29, S-42	GC, GC-MS, NMR, HPLC-DAD, MLCCC, MS, NMR, FAS, SDS, TLC	(Agarwal et al. 2001; Aburjai 2000; Krafczyk et al. 2008; Tayade et al. 2012; Banks and Cameron 1971; Wang et al. 2011)
Resveratrol	S-6, S-27, S-29, S-35, S-37, S-43	HPLC, MS	(Rokaya et al. 2012b; Piotrowska et al. 2012)
Rhaponticin-β-D-glucoside	S-6, S-37, S-43	NA	(Agarwal et al. 2001; Rehman et al. 2014)
Desoxyrhapontigenin	S-43	NA	(Agarwal et al. 2001; Wang et al. 2011)
Rhapontigenin	S-29, S-6, S-7, S-3, S-4, S-43	NA	(Rehman et al. 2014; Wang et al. 2011)
Piceatannol	S-5, S-6, S-43	NA	(Agarwal et al. 2001; Rehman et al. 2014)
4'-0-Methylpiceid	S-37	DCC	(Agarwal et al. 2001; Rehman et al. 2014)
Piceatannol-4'-0-β-D-glucopyranoside	S-5, S-6	NA	(Agarwal et al. 2001; Rehman et al. 2014)
Rhapontigenin-3'-0-β-D-glucopyranoside	S-38, S-6, S-43	NA	(Rehman et al. 2014; Wang et al. 2011)
Piceatannol-3'-0-β-D glucopyranoside (Rheumin)	S-38, S-42	GC, GC-MS, NMR, HPLC-DAD, MLCCC	(Krafczyk et al. 2008; Rehman et al. 2014)
Resveratrol-4-O-β-D-glucopyranoside	S-38, S-42, S-36, S-16,	GC, GC-MS, NMR, HPLC-DAD, MLCCC, MS, NMR	(Rokaya et al. 2012b; Krafczyk et al. 2008; Rehman et al. 2014; Aburjai 2000; Samappito et al. 2003),

(Continued)

Phytochemistry

TABLE 4.1 (CONTINUED)
Phytochemical Constituents Reported from Genus *Rheum*

Phytoconstituent	Species	Detection Method	Reference
3,5,4′-trihydroxystilbene 4′-O-β-D-glucopyranoside 6′-O-gallate	S-37, S-27	HPLC, DAD, ESI-MS, HPLC-MS, MS, CA, CDFP	(Jin et al. 2007; Lin et al. 2008)
Rhaponticin 2-0-gallate	S-45	H-NMR,13C NMR, CC, MS	(Matsuda et al. 2001)
Rhaponticin 6-O-gallate	S-45	H-NMR,13C NMR, CC, MS	(Matsuda et al. 2001)
Anthraglycosides			
Aloe-emodin-8-0-β-D-glucoside	S-37, S-27	LC-EIMS, HPLC, DAD, ESI-MS, HPLC- MS, MS	(Agarwal et al. 2001; Ye et al. 2007b; Jin et al. 2007)
Emodin-6-0-β-0-glucoside	S-37, S-27	HPLC, DAD, ESI-MS, HPLC-MS, MS	(Agarwal et al. 2001; Jin et al. 2007)
Rhein-8-0-β-D-glucoside	S-26, S-24, S-5, S-23, S-34, S-27, S-37, S-35, S-27	LC-EIMS, HPLC, DAD, ESI-MS, HPLC- MS, MS	(Jin et al. 2007; Ye et al. 2007b)
Rhein 1-O-glucoside	S-37, S-27	HPLC, DAD, ESI-MS, HPLC-MS, MS	(Jin et al. 2007; Ye et al. 2007b)
Rheinoside A	S-39, S-34, S-27, S-24, S-37, S-26	NA	(Agarwal et al. 2001)
Rheinoside B	S-6, S-7, S-9	NA	(Agarwal et al. 2001)
Rheinoside C	S-39, S-34, S-27, S-24, S-37, S-26, S-5,	NA	(Agarwal et al. 2001)
Rheinoside D	S-39, S-34, S-27, S-24, S-37, S-26, S-23, S-5	NA	(Agarwal et al. 2001)
Sennoside A	S-5, S-34, S-27, S-24, S-37, S-26, S-4, S-3, S-7, S-6	NA	(Agarwal et al. 2001; Tayade et al. 2012; Kemper and Research 1999)
Sennoside B	S-37, S-6, S-7, S-9	NA	(Agarwal et al. 2001)
Sennoside C	S-27, S-35, S-37,	HPLC, MS	(Ye et al. 2007b)

(Continued)

TABLE 4.1 (CONTINUED)
Phytochemical Constituents Reported from Genus *Rheum*

Phytoconstituent	Species	Detection Method	Reference
Sennoside D	S-27, S-35, S-37,	HPLC, MS	(Ye et al. 2007b)
Aloe-emodin-1-0-β-D-glucopyranoside	S-6, S-7, S-9	NA	(Agarwal et al. 2001)
Chrysophanein	S-37	NA	(Agarwal et al. 2001)
Emodin-O-glucoside	S-37, S-6, S-29, S-35, S-27	LC-EIMS	(Ye et al. 2007b)
Emodin 8-O-glucoside	S-37, S-6, S-29, S-35, S-27	LC-EIMS	(Ye et al. 2007b)
Emodin-8-0-β-D-glucopyranoside	S-27	HPLC, DAD, ESI-MS, HPLC-MS, MS	(Jin et al. 2007)
Chrysophanol 1-O-glucoside	S-37, S-6, S-29, S-35, S-27	LC-EIMS	(Ye et al. 2007b)
Chrysophanol-1-0-β-D-glucopyranoside	S-45	FAB-MS	(Matsuda et al. 2001)
Chrysophanol-8-0-β-D-glucopyranoside	S-45	FAB-MS	(Matsuda et al. 2001)
Chrysophanol-8-0-β-D--(6′-malonyl)-gluco pyranoside	S-27	HPLC, DAD, ESI-MS, HPLC-MS, MS	(Jin et al. 2007)
Physcion-8-0-β-D-glucoside	S-27	HPLC, DAD, ESI-MS, HPLC-MS, MS	(Jin et al. 2007)
Torachrysone 8-O-β-D glucopyranoside	S-45	H-NMR,13C NMR, CC, MS	(Matsuda et al. 2001)
6-Methyl rhein	S-6	HPLC	(Pandith et al. 2018)
6-Methyl aloe-emodin	S-6	HPLC	(Pandith et al. 2018)
Flavonoids			
Quercetin	S-2, S-5, S-16, S-38, S-42, S1	GC, GC-MS, NMR, HPLC-DAD, MLCCC	(Agarwal et al. 2001; Ragasa et al. 2017; Krafczyk et al. 2008)
Isoquercetin	S-5, S-16, S-2, S-38, S-42	GC, GC-MS, NMR, HPLC-DAD, MLCCC	(Agarwal et al. 2001; Ragasa et al. 2017; Krafczyk et al. 2008)
Quercetin-3, O-galactoside	S-1, S-40	R-HPLC, NMR, HPLC, PC	(Iwashina et al. 2004; Püssa et al. 2009)
Quercetin-3, O-rutinonoside	S-1, S-40	R-HPLC, NMR, HPLC, PC	(Iwashina et al. 2004; Püssa et al. 2009)
Meratin	S-16	GC, GC-MS, NMR, HPLC-DAD, MLCCC	(Agarwal et al. 2001; Krafczyk et al. 2008)

(Continued)

Phytochemistry

53

TABLE 4.1 (CONTINUED)
Phytochemical Constituents Reported from Genus *Rheum*

Phytoconstituent	Species	Detection Method	Reference
Rutin	S-5, S-16,	NA	(Agarwal et al. 2001)
Quercetin-3,7- glucoarbinoside	S-5, S-1, S-40	R-HPLC, NMR	(Agarwal et al. 2001; Iwashina et al. 2004; Püssa et al. 2009)
Quercetin 3-O-arabinopyranoside	S-1	UV, MS, NMR, HPLC, LC-MS	(Iwashina et al. 2004)
Quercetin 3-O-[6-(3-hydroxy-3-methy lglutaroyl)-glucoside]	S-1	UV, MS, NMR, HPLC, LC-MS	(Iwashina et al. 2004)
Kaempferol glycoside	S-1	UV, MS, NMR, HPLC, LC-MS	(Iwashina et al. 2004)
Quercetin 7-O-glycoside	S-1	UV, MS, NMR, HPLC, LC-MS	(Iwashina et al. 2004)
Feruloyl ester	S-1	UV, MS, NMR, HPLC, LC-MS	(Iwashina et al. 2004)
Myrecetin	S-6	HPLC	(Pandith et al. 2018)
Dianthrones			
Rheidin A	S-37, S-6, S-7, S-9	NA	(Agarwal et al. 2001)
Rheidin B	S-6, S-7, S-9	NA	(Agarwal et al. 2001)
Sennidin A	S-37	NA	(Abe et al. 2006b)
Sennidin B	S-37	NA	(Abe et al. 2006b)
Sennidin C	S-6, S-7, S-9, S-37	NA	(Agarwal et al. 2001; Abe et al. 2006b)
Rheidin C	S-6, S-7, S-9	NA	(Agarwal et al. 2001)
10-hydroxycascaroside C	S-6	NA	(Zargar et al. 2011)
10-hydroxycascaroside D	S-6	NA	(Zargar et al. 2011)
10-chrysaloin 1-O-b-D-glucopyranoside	S-6	NA	(Zargar et al. 2011)
Cascaroside C	S-6	NA	(Zargar et al. 2011)
Cascaroside D	S-6	NA	(Zargar et al. 2011)
Cassialoin	S-6	NA	(Zargar et al. 2011)
Polyphenols			
Carvacrol	S-44	HPLC, LC-MS	(Agarwal et al. 2001; Ye et al. 2007b)
Pyrogallol	S-44	NA	(Agarwal et al. 2001)
Kaempferol	S-4, S-6, S-7, S-9	LC-MS	(Agarwal et al. 2001; Ye et al. 2007b)
4-Methoxygallic acid	S-44	NA	(Agarwal et al. 2001)

(Continued)

TABLE 4.1 (CONTINUED)
Phytochemical Constituents Reported from Genus *Rheum*

Phytoconstituent	Species	Detection Method	Reference
(−)-Epicatechol	S-16, S-37	NA	(Agarwal et al. 2001)
(−)-Epicatechin-gallate	S-37, S-40, S-30, S-35, S-38, S-42	GC, GC-MS, NMR, HPLC-DAD, MLCCC	(Agarwal et al. 2001; Krafczyk et al. 2008)
Gallic acid	S-30, S-37, S-35	LC-MS	(Agarwal et al. 2001; Ye et al. 2007b)
Glucogallin	S-37, S-35, S-30, S-26	NA	(Agarwal et al. 2001)
1-O-galloyl-6-O-cinnamoyl-glucose	S-35	LC-MS	(Ye et al. 2007b)
(−)-Epigallocatechol	S-44	NA	(Agarwal et al. 2001)
(+)-Gallocatechol	S-44	LC-MS	(Agarwal et al. 2001)
(−)-Epigallocatechol-gallocatechol	S-44	NA	(Agarwal et al. 2001)
(+)-Catechol	S-37, S-16, S-5, S-4, S3, S-7	LC-MS	(Agarwal et al. 2001; Ye et al. 2007b; Tayade et al. 2012)
4.8′-bi s-3-0-Galloyl-(−)-epicatechin	S-4, S-6, S-7, S-9	HPLC	(Agarwal et al. 2001)
4.8′-3-0 -Galloyl-(−)-epicatechin	S-4, S-6, S-7, S-9	HPLC	(Agarwal et al. 2001)
(±)-Gallocatechol gallate	S-44	LC-MS	(Agarwal et al. 2001; Ye et al. 2007b)
(−)-Epicatechol -gallate	S-16, S-37	NA	(Agarwal et al. 2001)
P-Caumaric acid	S-37	HPLC, UV-MS	(Püssa et al. 2009)
b-Resorcylic acid	S-6	HPLC	(Pandith et al. 2018)
Daidzein-8-O-glucoside	S-6	HPLC	(Pandith et al. 2018)
Daidzein	S-6	HPLC	(Pandith et al. 2018)
(+)-Taxifolin	S-6	HPLC	(Pandith et al. 2018)
Organic acids			
Acetic acid	S-6, S-7, S-9	NA	(Agarwal et al. 2001)
Citric acid	S-6, S-7, S-9	NA	(Agarwal et al. 2001)
Fumaric acid	S-3, S-4, S-27	NA	(Agarwal et al. 2001)
Formic acid	S-3, S-4, S-27	NA	(Agarwal et al. 2001)
Palmitic acid	S-2, S-6, S-7, S-9	GC-GC/MS	(Ragasa et al. 2017)
Succinic acid	S-6, S-7, S-9	NA	(Agarwal et al. 2001)
Oxalic acid	S-7, S-3, S4	NA	(Agarwal et al. 2001; Tayade et al. 2012)

(Continued)

Phytochemistry

TABLE 4.1 (CONTINUED)
Phytochemical Constituents Reported from Genus *Rheum*

Phytoconstituent	Species	Detection Method	Reference
Lenoleic acid	S-2, S-6, S-7, S-9	GC-GC/MS	(Agarwal et al. 2001; Ragasa et al. 2017)
Malic acid	S-6, S-7, S-9	NA	(Agarwal et al. 2001)
Lactic acid	S-6, S-7, S-9	NA	(Agarwal et al. 2001)
GalLic acid	S-7, S-3, S4, S-27	HPLC-DAD, MLCCC, FAS, SDS	(Agarwal et al. 2001; Tayade et al. 2012)
Chromenes			
2,5-Dimethyl-7-hydroxy- chromene	S-37	NA	(Agarwal et al. 2001)
2-Methyl-5-acetyl-7-hydroxy-chromene	S-37	NA	(Agarwal et al. 2001)
2-2′-Hydroxylpropyl -5-methyl-7-hydroxychromene	S-37	NA	(Agarwal et al. 2001)
2-Methyl-5-carboxy methyl-7 -hydroxy-chromene	S-37	NA	(Agarwal et al. 2001)
2-Methyl-5-carboxymethyl-7-hydroxy chromanoe	S-37	NA	(Agarwal et al. 2001)
2-Methyl-5-(-2′-oxo-4-hydroxypentyl)-7-hydroxychromone-7 -0-β-D)-glucopyranoside	S-37	NA	(Agarwal et al. 2001)
Aloesone-7-0-β-D glucopyranoside	S-37	NA	(Agarwal et al. 2001)
Essential oils			
Paeonol	S-37	NA	(Agarwal et al. 2001)
Eugenol	S-6	NA	(Agarwal et al. 2001)
Methyl heptyl ketone	S-6	NA	(Agarwal et al. 2001)
P-Cadinene	S-37	NA	(Agarwal et al. 2001)
α-Copaene	S-37	NA	(Agarwal et al. 2001)
Methyl-eugenol	S-37	NA	(Agarwal et al. 2001)
Methyl stearate	S-37	NA	(Agarwal et al. 2001)
Brassinosteroids			
Castasterone	S-38	GC-MS, DEA, HPLC, H-NMR, C-NMR	(Schmidt et al. 1995)
Epicastesterone	S-38	DEA, HPLC, H-NMR, C-NMR	(Schmidt et al. 1995)
Campestrol	S-38	GC, GC-MS	(Schmidt et al. 1995)
Stigmastrol	S-38	GC, GC-MS	(Schmidt et al. 1995)

(Continued)

TABLE 4.1 (CONTINUED)
Phytochemical Constituents Reported from Genus *Rheum*

Phytoconstituent	Species	Detection Method	Reference
Isofucosterol	S-38	GC, GC-MS	(Schmidt et al. 1995)
Phenylbutanoids			
lindleyin	S-37, S-27	CA, CDFP, HPLC, DAD, ESI-MS, HPLC-MS, MS	(Lin et al. 2008; Jin et al. 2007)
Isolindleyin	S-37, S-27	CA, CDFP, HPLC, DAD, ESI-MS, HPLC-MS, MS	(Lin et al. 2008; Jin et al. 2007)
4-(4'-hydroxyphenyl)-2-butanone 4'-O-β-D-glucopyranoside	S-37, S-27	CA, CDFP, HPLC, DAD, ESI-MS, HPLC-MS, MS	(Lin et al. 2008; Jin et al. 2007)
4-(4'-hydroxyphenyl)-2-butanone	S-37, S-27	CA, CDFP, HPLC, DAD, ESI-MS, HPLC-MS, MS	(Lin et al. 2008; Jin et al. 2007)
Anthocyanins			
Cyanidin-3-glucoside	S-6, S-7, S-9	NA	(Agarwal et al. 2001)
Chrysanthemin	S-6, S-7, S-9	NA	(Agarwal et al. 2001)
Cyanidin-3-rutinoside	S-16	NA	(Agarwal et al. 2001)
Cyanin	S-16	NA	(Agarwal et al. 2001)
Phenylglycosides			
Gallic acid-4-O-β-D-pyranoside	S-27	HPLC, DAD, ESI-MS, HPLC-MS, MS	(Jin et al. 2007)
Gallic acid-3-O-β-D-pyranoside	S-27	HPLC, DAD, ESI-MS, HPLC-MS, MS	(Jin et al. 2007)
Esters and ethers			
Revandchinone-1	S-6	MS, NMR, IR	(Babu et al. 2003)
Revandchinone-2	S-6	MS, NMR, IR	(Babu et al. 2003)
Revandchinone-4 (anthraquinone ether)	S-6	MS, NMR, IR	(Babu et al. 2003)
Revandchinone-3 (oxanthrone ether)	S-6	MS, NMR, IR	(Babu et al. 2003)
Ferulic ester	S-37	HPLC, UV-MS	(Püssa et al. 2009)
Naphthalenes			
Torachrysone 8-O-glucoside	S-6, S-27, S-29, S-35, S-37, S-43	HPLC, MS	(Ye et al. 2007b)
Torachrysone 8-O-(6-Oacetyl)-glucoside	S-6, S-27, S-29, S-35, S-37, S-43	HPLC, MS	(Ye et al. 2007b)

(Continued)

Phytochemistry   57

TABLE 4.1 (CONTINUED)
Phytochemical Constituents Reported from Genus *Rheum*

Phytoconstituent	Species	Detection Method	Reference
Vitamins			
Vitamin B	S-4, S-6, S-7, S-9	NA	(Agarwal et al. 2001)
Vitamin C	S-4, S-6, S-7, S-9	NA	(Agarwal et al. 2001)
Vitamin D	S-4, S-6, S-7, S-9	NA	(Agarwal et al. 2001)
Other compounds			
Chlorophyllide	S-2	TLC, NMR	(Ragasa et al. 2017)
Methyl butanol	S-4, S-3, S-7	GC, MD/GC	(Dregus et al. 2003; Tayade et al. 2012)
2-methyl butanol	S-38	GC, MD/GC	(Dregus et al. 2003)
2-methyl butanoic acid	S-38	GC, MD/GC	(Dregus et al. 2003)
4-methyl hexanoic acid	S-38	GC, MD/GC	(Dregus et al. 2003)
Pulmatin	S-37	DCC, R-HPLC, NMR	(Agarwal et al. 2001)
Carpusin	S-6	NA	(Zargar et al. 2011)
Maesopsin	S-6	NA	(Zargar et al. 2011)
sulfemodin 8-O-b-D-glucoside	S-6	NA	(Zargar et al. 2011)

R. nobile (**S-1**), *R. ribes* (**S-2**), *R. tibeticum* (**S-3**), *R. moorcroftianum* (**S-4**), *R. wittrockii* (**S-5**), *R. australe* (**S-6**), *R. webbianum* (**S-7**), *R. globulosum* (**S-8**), *R. spiclforme* (**S-9**), *R. reticulatum* (**S-10**), *R. maculatum* (**S-11**), *R. compactum* (**S-12**), *R. przewalskyi* (**S-13**), *R. rhizostachyum* (**S-14**), *R. rhomboideum* (**S-15**), *R. tataricum* (**S-16**), *R. uninerve* (**S-17**), *R. nanum* (**S-18**), *R. sublanceolatum* (**S-19**), *R. racemiferum* (**S-20**), *R. inopinatum* (**S-21**), *R. subacaule* (**S-22**), *R. pumilum* (**S-23**), *R. delavayi* (**S-24**), *R. yunnanense* (**S-25**), *R. kialense* (**S-26**), *R. tanguticum* (**S-27**), *R. laciniatum* (**S-28**), *R. hotaoense* (**S-29**), *R. altaicum* (**S-30**), *R. glabricaule* (**S-31**), *R. forrestii* (**S-32**), *R. likiangense* (**S-33**), *R. lhasaense* (**S-34**), *R. officinale* (**S-35**) (**S-41**), *R. palastenium* (**S-36**), *R. palmatum* (**S-37**), *R. rhabarbamm* (**S-38**), *R. alexandrae* (**S-39**), *R. acuminatum* (**S-40**), *R. rhaponticum* (**S-42**), *R. franzenbachii* (**S-43**), *R. maximoviczii* (**S-44**), *R. undulatum* (**S-45**)

into several classes including flavones, flavonols, flavanones, isoflavones, and anthocyanin pigments (Pandith et al. 2016). In general, flavonols are the abundant group of all flavonoids which typically accumulate as glycosides of kaempferol, quercetin, or myricetin in the vacuoles of plant cells. Evidently, these (flavonols) form important ingredients in our diet (as humans/animals are unable to synthesize flavonoids *in situ*) and have also been reported to act as vital agents against some of the prominent ill effects of human health (Stafford 1990; Pandey et al. 2015; Pandith et al. 2016). In plants, flavonoids play diverse roles in many aspects including auxin transport, cell wall growth, pollen development, flower coloration, photoprotection, and response to stress conditions like herbivory, UV light protection, wounding, defense against pathogens, and interaction with soil microbes (Pandey et al. 2015; Pandith et al. 2016). A series of biological studies on *R. nobile* has revealed a greenhouse effect of the translucent bracts for the inflorescences against extreme cold in high mountains, like the Himalayas. It is presumed that the epidermis and hypodermal layers of *R. nobile* contain many flavonoids dissolved in their cells. Moreover, these bracts are presumed to provide defense against noxious UV radiation, especially UV-B; these translucent bracts screen UV radiation and only allow visible light to pass (Iwashina et al. 2004). In genus *Rheum* more than a dozen flavonoids have been characterized so far including quercetin, iso-quercetin, meratin, rutin, quercetin galactosides, quercetin rutinonosides, quercetin arabinopyranosides, quercetin glucoarbinosides, and kaempferol glycosides (Agarwal et al. 2001). Structures of some of the representative phenolics are provided in Figure 4.1.

The flavonoid biosynthesis mechanism has remained one of the most intensively studied metabolic systems in plants. The pathway has been well characterized in *Arabidopsis thaliana* and to a large extent in *Vitis vinifera* and *Zea may* (Castellarin and Di Gaspero 2007; Boss et al. 1996; Bogs et al. 2006). Functional genomic studies, particularly transcriptomic and metabolomic approaches, have largely been used to extensively characterize the genes involved in the biosynthesis of flavonoids and their subsequent modified moieties. In Polygonaceae, like in other plant families, flavonoids are synthesized by combination of the polyketide pathway and phenylpropanoid pathway. The later provides the p-coumaroyl CoA from phenylalanine (Phe) and tyrosine (Tyr), whereas the former caries out chain elongation by utilizing the extender

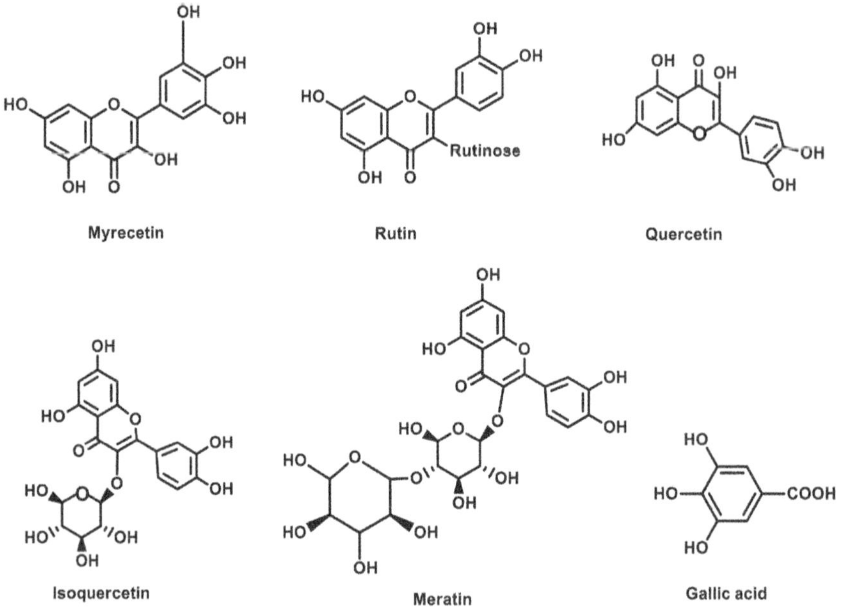

FIGURE 4.1 Representative phenolic constituents isolated from genus *Rheum*.

(Continued)

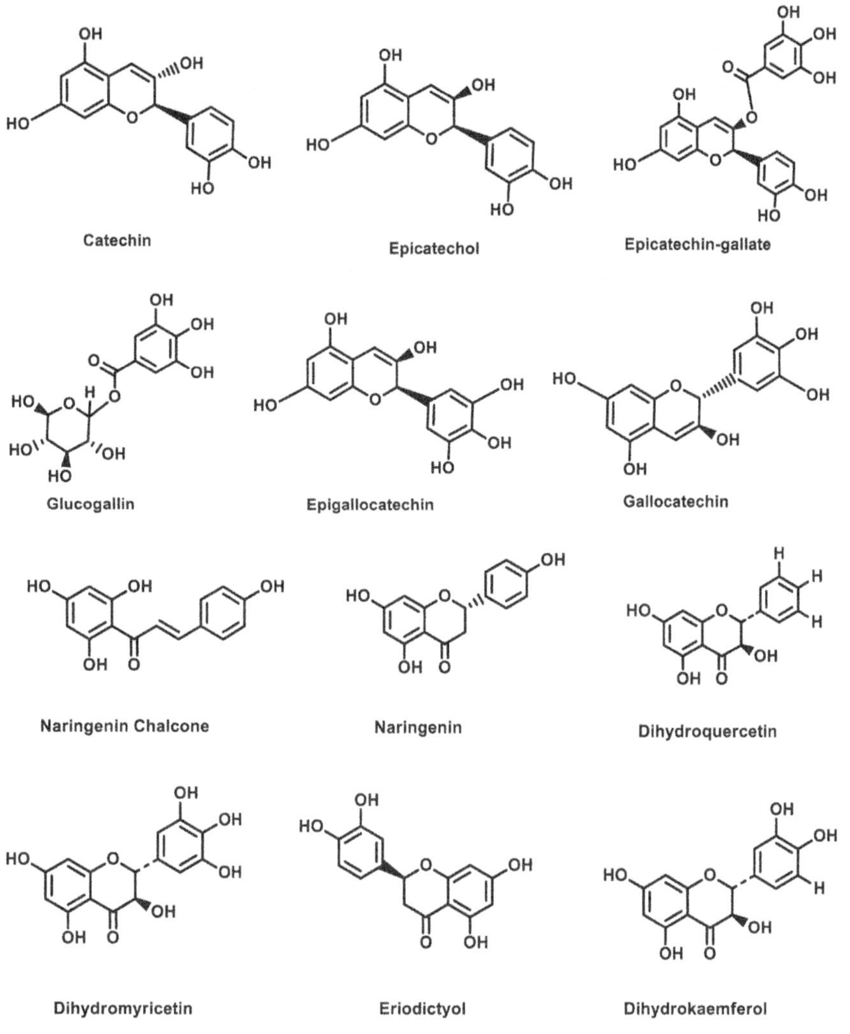

Phytochemistry

59

FIGURE 4.1 (Continued) Representative phenolic constituents isolated from genus *Rheum*.

malonyl-CoA molecule. The aromatic amino acids Phe and Tyr are synthesized from the shikimate pathway which has been well explored in *A. thaliana* (Fraser and Chapple 2011; Maeda and Dudareva 2012). In most species three enzymes are required to transform phenylalanine into CoA ester which then enters the downstream pathway. The first committed enzyme in the phenylpropanoid pathway phenylalanine ammonia lyase (PAL) catalyzes the deamination of phenylalanine to produce trans-cinnamic acid from phenylalanine (Calabrese et al. 2004; MacDonald and D'Cunha 2007). Second, cinnamic acid 4-hydroxylase (C4H), a cytochrome P450 hydroxylase catalyzes the introduction of the hydroxyl group at para position of the phenyl ring of trans-cinnamic acid to yield p-coumaric acid (4-coumaric acid) (Munro et al. 2007; Rosler et al. 1997). The final thioester activation step is catalyzed by the enzyme coumaric acid: CoA ligase (4CL) which involves the activation of 4-coumaric acid by forming a thioester bond with CoA to produce p-coumaroyl CoA (Dixon and Paiva 1995; Ferrer et al. 1999). On the other hand, the condensing unit from the polyketide pathway malonyl-CoA is synthesized by the carboxylation of acetyl-CoA, a central intermediate in tricarboxylic acid (TCA) cycle. The reaction is catalyzed by a Mg^{2+}-ATP-dependent biotinylated protein complex ACC (acetyl-CoA carboxylase) (Nikolau et al. 2003; Alban et al. 2000; Sasaki and Nagano 2004). Chalcone synthase (CHS) is the first committed enzyme in the central phenylpropanoid biosynthetic pathway which catalyzes claisen condensation of three malonyl-CoA molecules and one molecule of p-coumaroyl CoA to produce a polyketide intermediate that undergoes further condensation, cyclization, and isomerization events to generate naringenin (Austin and Noel 2003). Various downstream enzymes act to result in the formation of various other products, viz., dihydrokaempferol, dihydroquercetin, dihydromyricetin, eriodictyol, quercetin, kaempferol, etc. The final biological and physicochemical properties of the flavonoid scaffold structures generated from the central phenylpropanoid pathway are determined by the type of modification by different tailoring enzymes such as methyltransferases, glycosyltransferases, and acyltransferases (Pandith et al. 2018).

Additionally, 17 different types of polyphenols have been reported in the genus *Rheum* so far. The different polyphenols found in rhubarb include glucogallin, (–)-epicatechol, (–)-epicatechin-gallate, (±)-gallocatechol gallate, (–)-epicatechol -gallate, (+)-gallocatechol, carvacrol, pyrogallol, and kaempferol, etc. The polyphenols from genus *Rheum* consist chiefly of gallic acid which is present as glucogallin and catechin with small amounts of tannin. The glucogallin is the principal polyphenol found in rhubarb which upon hydrolysis yields gallic acid and glucose (Agarwal et al. 2001). The galloyl esters of glucose, such as glucogallin, are an important class of phenolic compounds found in rhubarb that show antitumor, anticancer, and antidiabetic potential. The glucose gallates have been found from both official and unofficial varieties of rhubarb. All glucogallins contain at least one galloyl group on the glucose present on the C-1, C-2, or C-6 atom (Ye et al. 2007b). Further, and as stated earlier, catechin with gallic acid forms most of the polyphenols in rhubarb; catechin *per se* as a free compound falls in one of the 12 classes of flavonoids. It contains hydroxy group at C-3 of the central pyran ring. Some 70 catechins are known to occur in different species of plants such as (2R,2S)-3,3',4',5,7-flavanpentol among others. Catechins are widely distributed in plants arising biosynthetically from the shikimate pathway and are said to be responsible for the antitumor and anticancer properties of polyphenols in rhubarb (Cammack et al. 2000). Polyphenols are also shown to block the formation of endothelin-1, a signaling molecule that constricts blood vessels. This effect of tannins accounts for their health benefits, especially the reduction in the risk of heart disease (Corder et al. 2001). A detailed account of phenols reported from rhubarb is given in Table 4.1.

Phytochemistry 61

4.2 ANTHRAQUINONES

Anthraquinones (anthracenediones or dioxoanthracenes; $C_{14}H_8O_2$) are an imperative and varied class of aromatic organic compounds which show their presence across living systems, viz., higher plants, fungi, lichens, and even bacteria. To date, about 200 natural anthraquinones have been recorded from fungi, lichens, and some of the major plant families, viz., Xanthorrhoeaceae, Rubiaceae, Rhamnaceae, and Polygonaceae (Singh et al. 2004; Pandith et al. 2018; Chien et al. 2015). Mass spectrometric analyses have deciphered the identity of about 107 phenolic compounds from the genus *Rheum* which mainly include anthraquinones and stilbenoids, as well as naphthalenes, catechins, sennosides, and glucose gallates (Ye et al. 2007a; Agarwal et al. 2001). The anthraquinones can be broadly categorized into two types on the basis of their structural differences and the biosynthetic routes. The structural differences arise as a result of different hydroxylation patterns accompanied by different biosynthetic pathways. One class of compounds is biosynthesized through an acetate/malonate pathway (polyketide pathway) in which the rings A and C both are hydroxylated. The other class of compounds is biosynthesized through a succinyl-benzoate pathway in which only ring C is hydroxylated. Anthraquinones are not only found in their free form in different species of plants; instead, many of them are linked to the sugar moieties to form the glycosides (Pandith et al. 2018). Besides flavonoids being the potent phytoceuticals for human health, some of the major anthraquinone constituents mostly isolated from *R. australe* have been shown to display a wide pharmacological significance. A detailed investigation of all such medical efficacies was done earlier by us, a comprehensive report of which can be found in our previous communication (Pandith et al. 2018). Structures of some of the representative anthraquinones are provided in Figure 4.2.

The anthraquinones in higher plant species are synthesized through two different and key biosynthetic pathways, viz., the acetyl/polymalonyl polyketide pathway and the chorismate/o-succinyl benzoic acid pathway. The former occurs in fungi and some higher plant families like Rhamnaceae, Polygonaceae, and Fabaceae (Pandith et al. 2018) and leads to the biosynthesis of anthraquinones, naphthoquinones, and several phenylpropanoids. In this pathway, anthraquinones are synthesized from CoA esters through different condensation and cyclization events carried out by specific promiscuously behaving enzymes called polyketide synthases (PKSs) which indeed form a major gene family in bacteria, fungi, and plants as well. The condensation of one acetyl-CoA unit and seven malonyl-CoA units via an octaketide intermediate lead to the formation of anthraquinones (Figure 4.3). The chorismate/o-succinylbenzoic acid pathway is mainly found in the members of Rubiaceae family and the anthraquinones that are synthesized are commonly called *Rubia*-type anthraquinones (Pandith et al. 2018). In fact, Cinchona (Han et al. 2002), *Galium* (Inoue et al. 1984; Bauch and Leistner 1978), Morinda (Leistner 1974), and *Rubia* (Leistner 1981) belonging to the Rubiaceae family have been extensively used to study the biosynthesis of *Rubia*-type anthraquinones. Earlier studies involving feeding experiments have established that

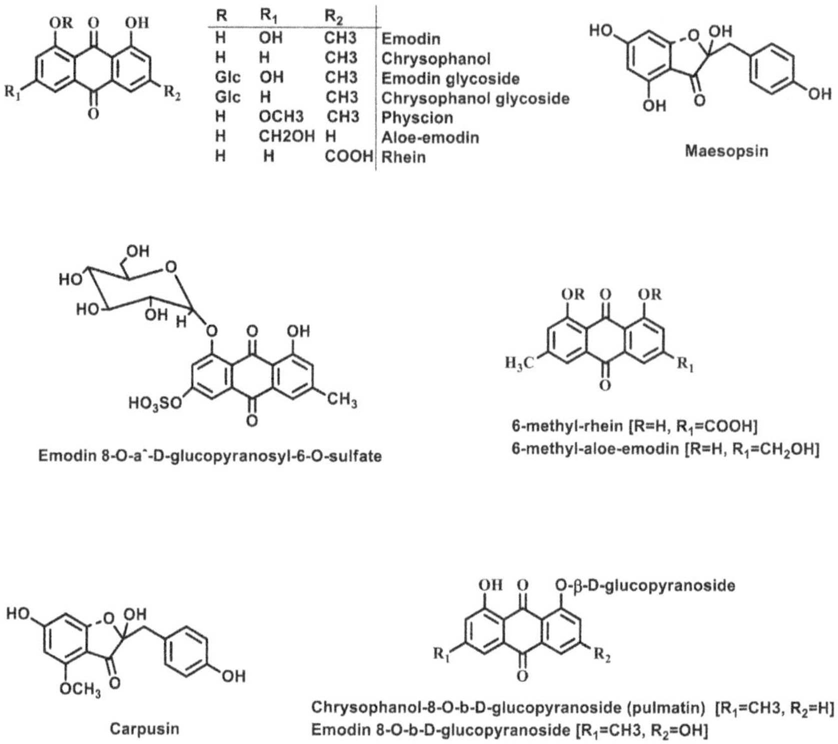

FIGURE 4.2 Representative anthraquinone constituents isolated from genus *Rheum*.

rings A and B of *Rubia*-type anthraquinones are biosynthetically derived from shikimate, α-ketoglutarate via o-succinylbenzoate (Han et al. 2001). Moreover, recent findings have demonstrated that the isopentenyl diphosphate (IPP) moiety of ring C comes from non-mevalonate (MEP) and not the mevalonate (MVA) pathway (Han et al. 2002). At present, relatively few genes from this pathway have been partially characterized to the enzyme level. However, the enzymatic steps leading to the biosynthesis of *Rubia*-type anthraquinones are still not completely understood (Han et al. 2001). Further, and unlike the biosynthesis of these anthraquinones, coumarins and flavonoids are synthesized from various precursors involving distinct biosynthetic routes (Leistner 1995).

4.3 STILBENOIDS

Stilbenoids (C_6–C_2–C_6) are the hydroxylated derivatives of stilbenes which belong to the phenylpropanoid family. Stilbenes are actually the low-molecular-weight (~200–300 g/mol), naturally occurring organic compounds that are found in a wide range of plant sources, dietary supplements, and aromatherapy products. They are synthesized via the phenylpropanoid pathway which is activated upon environmental threat leading to subsequent production and secretion of these

Phytochemistry

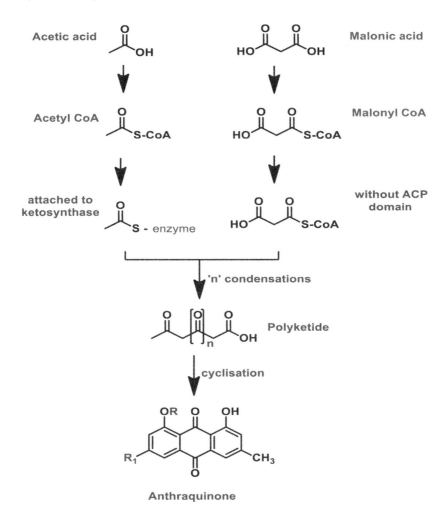

FIGURE 4.3 *The chemistry of polyketide chain assembly:* Acetic acid and malonic acid are converted to their coenzyme A (CoA) esters and then attached, by specific acyl transferases, to components of the polyketide synthase (PKS); acetyl-CoA is attached to the active site of the ketosynthase, and malonyl-CoA to a structural component of the PKS called the acyl carrier protein (ACP), usually absent in type III PKSs. Condensation of the two units by the ketosynthase, with loss of one carbon from malonyl-CoA as carbon dioxide, produces a four-carbon chain attached to the ACP. This is transferred back to the ketosynthase, and further rounds of condensation with malonyl-CoA or other chain extender units produce a polyketide chain which then cyclizes to anthraquinones.

molecules. Stilbenes protect the host plant against excessive ultraviolet exposure, and from different viral and microbial attacks (Roupe et al. 2006). So far, stilbene compounds have been isolated from more than 70 different plant species including some edible plants, such as, peanuts, grapes and various berries (Rokaya et al. 2012a).

Besides anthraquinones, rhubarb also serves as a rich repository of stilbenes, viz., rhaponticin, piceatannol, deoxyrhaponticin, rhaponticin-β-D-glucoside, resveratrol, rhapontigenin, deoxyrhapontigenin, etc. (Agarwal et al. 2001). Moreover, the stilbenes are known to share their biosynthetic pathway with chalcones. For instance, the biosynthesis of stilbenes is carried out by the stilbene synthase (STS) enzyme which incurs the same substrates as CHS. In fact, the two enzymes are considered to have evolved independently in the evolutionary course (Pandith et al. 2018). Rhubarb is a rich and valuable source of the major stilbenoid compound, resveratrol. Even under cultivated conditions, the concentration of the compound is equivalent to those in different types of wines and peanuts (Rokaya et al. 2012a). Piceatannol (3, 30, 4, 50-trans-trihydroxystilbene) is a less studied stilbenoid than resveratrol. But piceatannol is a naturally occurring hydroxylated analog of resveratrol. It is found in various plants, including passion fruit, grapes, Japanese knotweed, and white tea and displays a wide spectrum of biological activity (Piotrowska et al. 2012). Structures of some of the representative stilbenoids are shown in Figure 4.4.

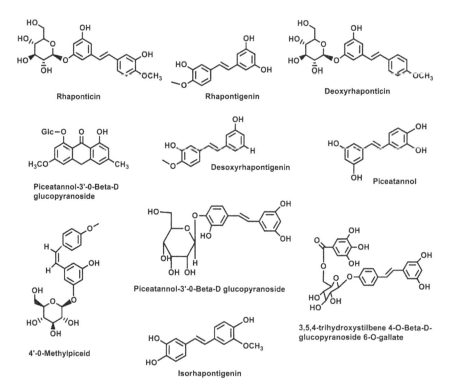

FIGURE 4.4 Representative stilbenoid compounds isolated from genus *Rheum*.

(*Continued*)

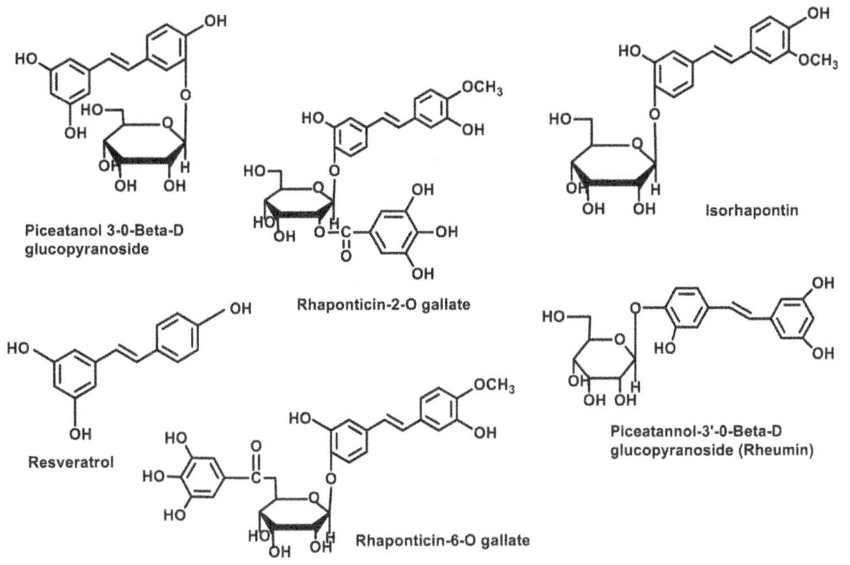

Phytochemistry 65

FIGURE 4.4 (Continued) Representative stilbenoid compounds isolated from genus *Rheum*.

4.4 OTHER CONSTITUENTS

In addition to the class of compounds discussed above, rhubarb is known to harbor various other chemical constituents with or without pharmacological significance. These include anthraglycosides, dianthrones, brassinosteroids, anthocyanins, ethers, esters, and naphthalenes, etc. A detailed account of all these compounds is provided in Table 4.1, and structures of some representatives from these classes are given in Figure 4.5.

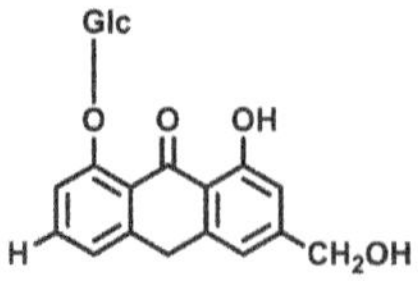

Aloe-emodin- 8-0-beta-D-glucopyranoside

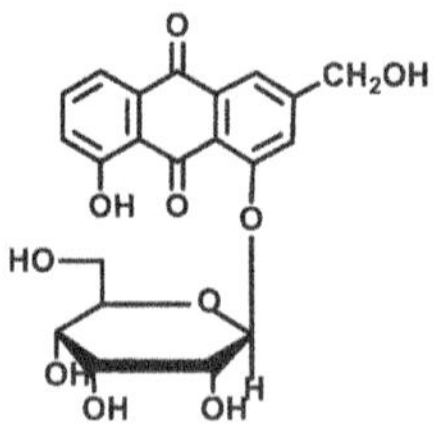

Aloe-emodin- 1-0-beta-D-glucopyranoside

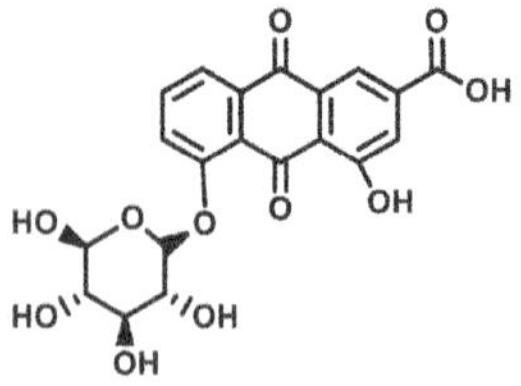

Rhein 1-O-glucoside

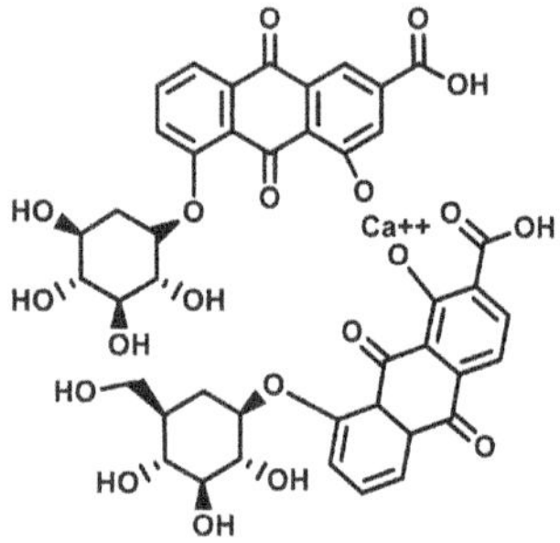

Rhein-8-0-beta-D-glucoside

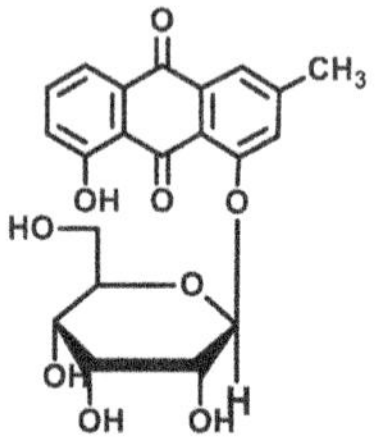

Chrysophanol 8-0-beta-D-glucopyranoside

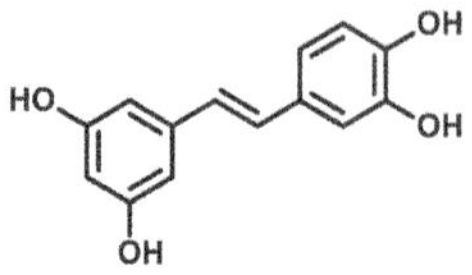

Piceatannol 4-O-beta-D Glucopyranoside

FIGURE 4.5 Representative anthraglycosides, dianthrones, brassinosteroids, anthocyanins, ethers, esters, organic acids, and naphthalenes, etc. isolated from genus *Rheum*.

(*Continued*)

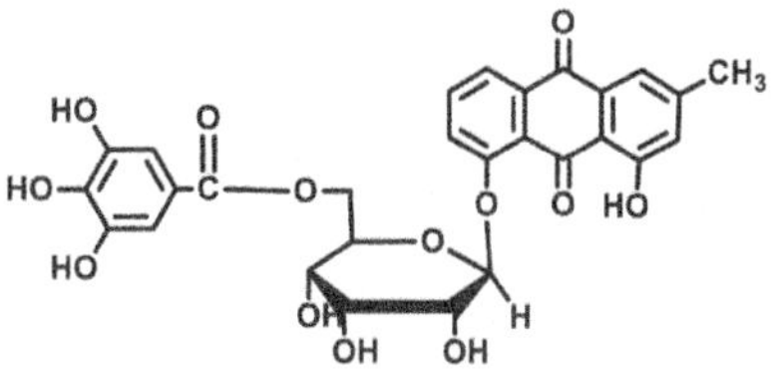

Chrysophanol-8-0-Beta_D--
(6'-malonyl)-glucopyranoside

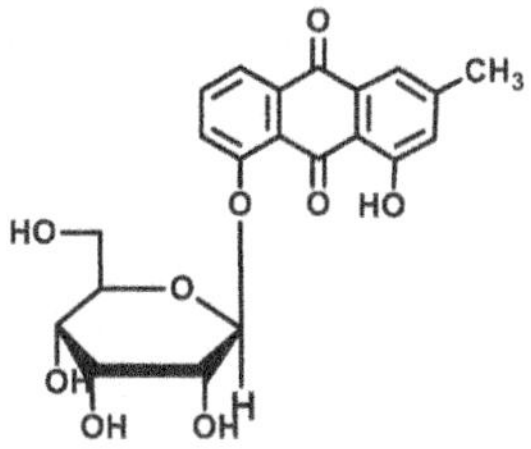

Chrysophanol 1-0-Beta-
D-glucopyranoside

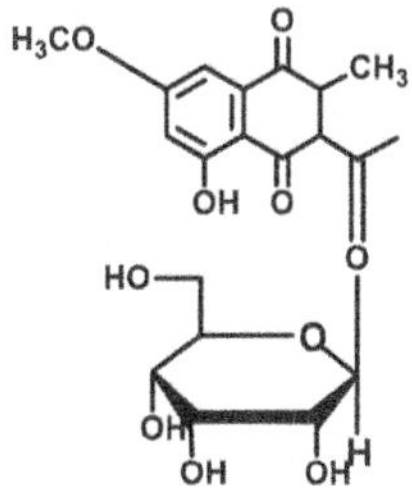

Torachrysone 8-0-Beta-
D-glucopyranoside

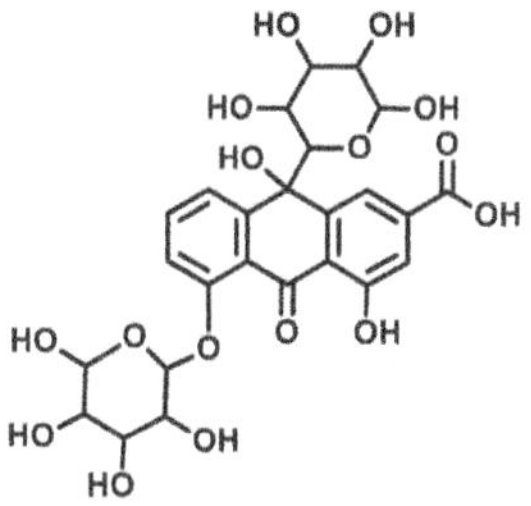

Rheinoside A

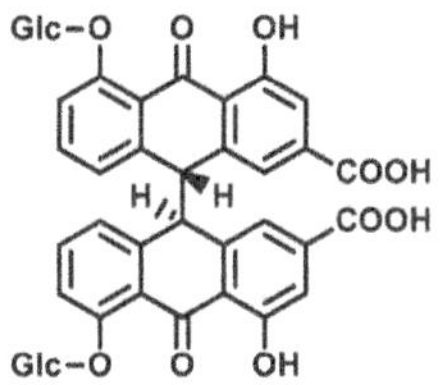

Sennoside A

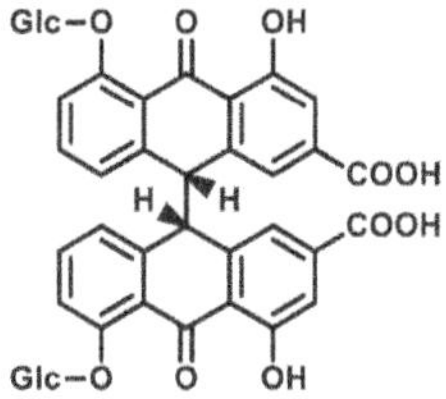

Sennoside B

FIGURE 4.5 (Continued) Representative anthraglycosides, dianthrones, brassinosteroids, anthocyanins, ethers, esters, organic acids, and naphthalenes, etc. isolated from genus *Rheum*.

(Continued)

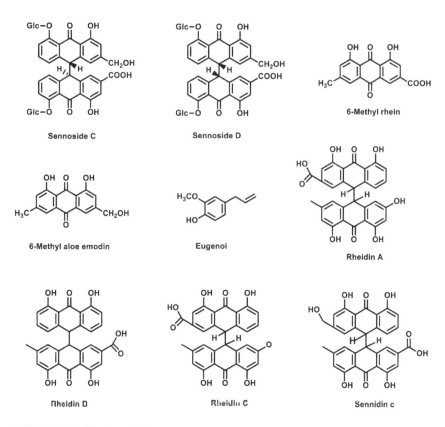

FIGURE 4.5 (Continued) Representative anthraglycosides, dianthrones, brassinosteroids, anthocyanins, ethers, esters, organic acids, and naphthalenes, etc. isolated from genus *Rheum*.

(*Continued*)

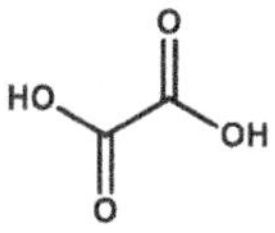

Oxalic acid

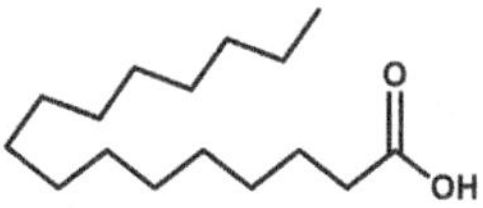

Palmitic acid

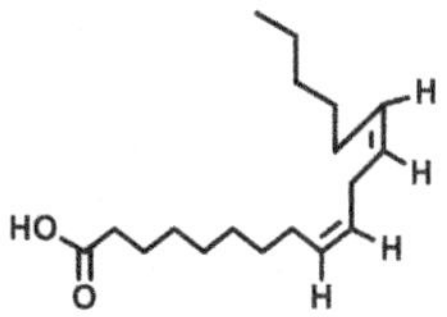

Lenoleic acid

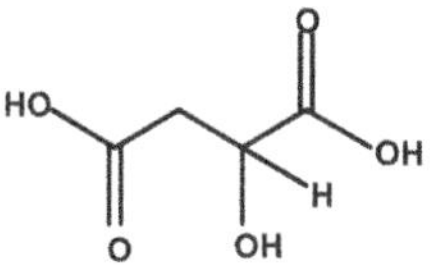

Malic acid

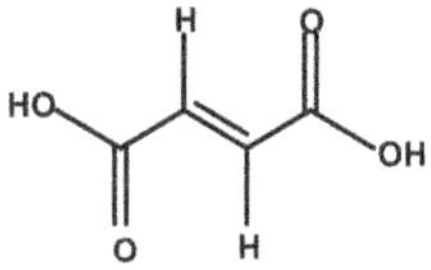

Fumaric acid

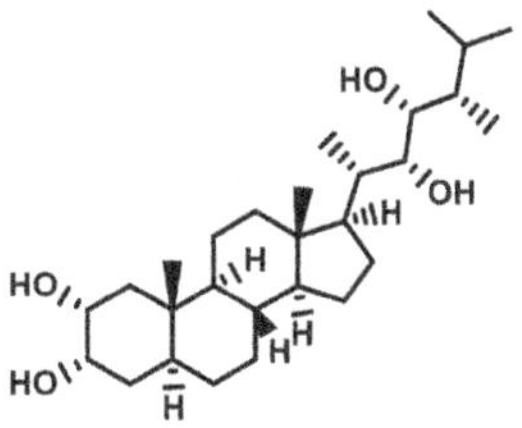

Castasterone

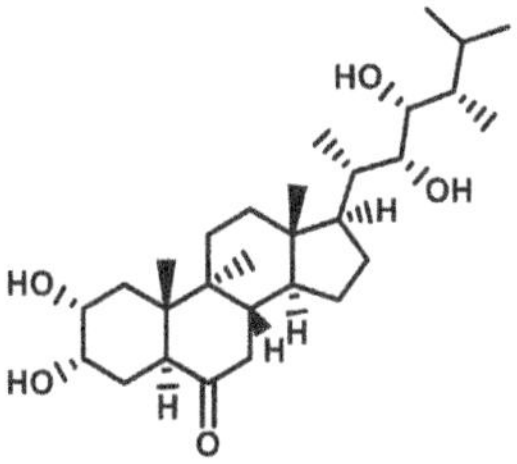

Epicastesterone

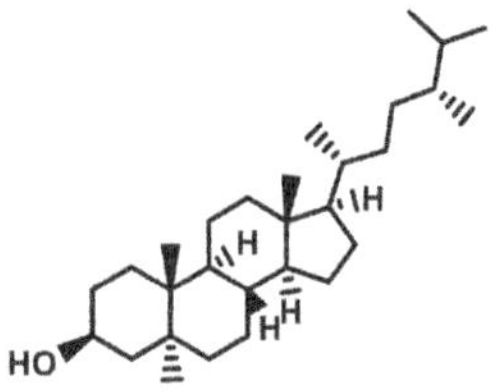

Campestrol

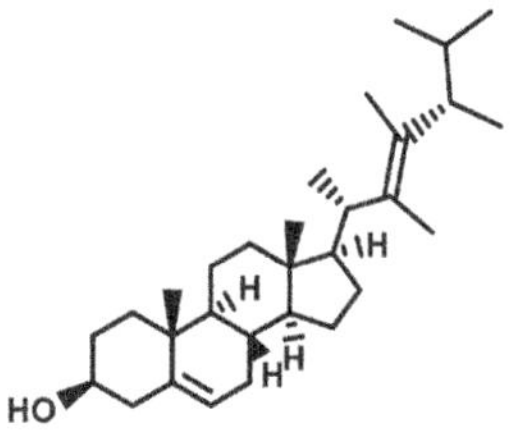

Stigmastrol

FIGURE 4.5 (Continued) Representative anthraglycosides, dianthrones, brassinosteroids, anthocyanins, ethers, esters, organic acids, and naphthalenes, etc. isolated from genus *Rheum*.

(Continued)

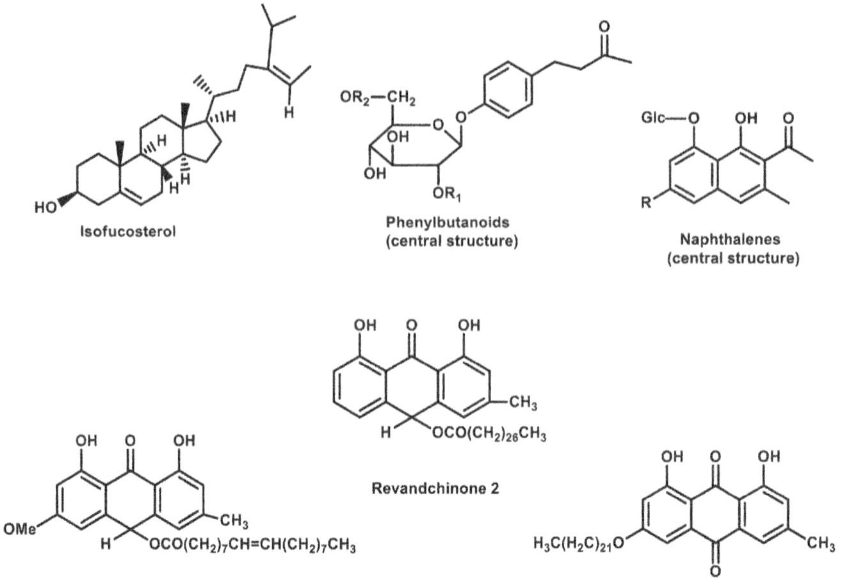

The figure shows chemical structures: Isofucosterol, Phenylbutanoids (central structure), Naphthalenes (central structure), Revandchinone 2, Revandchinone 1, Revandchinone 3.

FIGURE 4.5 (Continued) Representative anthraglycosides, dianthrones, brassinosteroids, anthocyanins, ethers, esters, organic acids, and naphthalenes, etc. isolated from genus *Rheum*.

4.4.1 ANTHRAGLYCOSIDES

To date, about 21 anthraglycosides have been characterized in different species of rhubarb including emodin glucosides, emodin glucopyranosides, aloe emodin glucosides, aloe-emodin glucopyranosides, rhein glucosides, rheinosides, sennosides, chrysophan glucosides, physcion glucosides, and torachrysone glucopyranosides. These different anthraglycosides are found in almost all the species of rhubarb (Agarwal et al. 2001). In *R. palmatum*, the oriental medical plant, the sennosides A and B are the main constituents of its rhizome. These sennosides, being themselves inactive, are activated by intestinal bacteria in the gut. They are first hydrolyzed to sennidin monoglucosides and then to sennidin A and B by β-glucosidase. The sennidin A and B are finally reduced to rhein anthrone which is considered to exert the laxative effect on the colon (Abe et al. 2006a). Moreover, sennoside A is considered as the major purgative compound of rhubarb. Along with rhein 8-O-glucoside and emodin 1-O-glucoside, it is mainly found in *R. officinale* (southern rhubarb) in high levels. Similarly, rhein 1-O-glucoside and emodin 8-O-glucoside are found in *R. palmatum* and *R. tanguticum* (northern rhubarb) (Ye et al. 2007b). A detailed account of anthraglycosides reported from rhubarb is given in Table 4.1.

Phytochemistry

4.4.2 DIANTHRONES

Dianthrones are an important group of chemical compounds found in different plant species including *Rheum*. These are actually the chemical derivatives of anthraquinones. To date, six dianthrone compounds have been reported in rhubarb including rheidin A, B, C, and sennidin A, B (a stereoisomer of sennidin A), and C. Rheidin A consists of one mole of emodin and one mole of rhein, rheidin B consists of one mole of rhein and one mole of chrysophanol, and rheidin C consists of one mole of rhein and one mole of physcion. Similarly sennidin C consists of one mole of aloe-emodin and one mole of rhein (Abe et al. 2006a; Agarwal et al. 2001).

Abe et al. (2006a) have found that sennidin A stimulates glucose incorporation in rat adipocytes near to the maximal level. Sennidin B was also shown to stimulate glucose incorporation, but the activity of sennidin B was lower than sennidin A. It has been shown that sennidin A induced both Thr-308 and Ser-473 phosphorylation of Akt, following activation of PI3K (phosphatidylinositol 3-kinase) and glucose transporter 4 (GLUT4) translocation (Abe et al. 2006a). A detailed account of dianthrones reported from rhubarb is given in Table 4.1.

REFERENCES

Abe D, Saito T (2006a) Sekiya KJLs. *Sennidin Stimulates Glucose Incorporation in Rat Adipocytes* 79(11):1027–1033.

Abe T, Noma H, Noguchi H (2006b) Enzymatic formation of an unnatural methylated triketide by plant type III polyketide synthases. *Tetrahedron Letters* 47(49):8727–8730.

Abrol E, Vyas D, Koul S (2012) Metabolic shift from secondary metabolite production to induction of anti-oxidative enzymes during NaCl stress in Swertia chirata Buch.-Ham. *Acta Physiologiae Plantarum* 34(2):541–546.

Aburjai TAJP (2000) Anti-platelet stilbenes from aerial parts of Rheum palaestinum. *Phytochemistry* 55(5):407–410.

Agarwal SK, Singh SS, Lakshmi V, Verma S, Kumar S (2001) *Chemistry and Pharmacology of Rhubarb (Rheum species)—A Review*. New Delhi, India: NISCAIR-CSIR.

Alban C, Job D, Douce R (2000) Biotin metabolism in plants. *Annual Review of Plant Biology* 51(1):17–47.

Austin MB, Noel JP (2003) The chalcone synthase superfamily of type III polyketide synthases. *Natural Product Reports* 20(1):79–110.

Babu KS, Srinivas P, Praveen B, Kishore KH, Murty US, Rao JMJP (2003) Antimicrobial constituents from the rhizomes of Rheum emodi. *Phytochemistry* 62(2):203–207.

Banks H, Cameron DJA (1971) JoC. New Natural Stilbene Glucoside from Rheum Rhaponticum (Polygonaceae). *Australian Journal of Chemistry* 24(11):2427–2430.

Bauch H, Leistner E (1978) Aromatic Metabolites in Cell Suspension Cultures of Galium mollugo L. *Planta Medica* 33:105–123.

Bogs J, Ebadi A, McDavid D, Robinson SP (2006) Identification of the flavonoid hydroxylases from grapevine and their regulation during fruit development. *Plant Physiology* 140(1):279–291.

Boss PK, Davies C, Robinson SP (1996) Analysis of the expression of anthocyanin pathway genes in developing Vitis vinifera L. cv Shiraz grape berries and the implications for pathway regulation. *Plant Physiology* 111(4):1059–1066.

Calabrese JC, Jordan DB, Boodhoo A, Sariaslani S, Vannelli T (2004) Crystal structure of phenylalanine ammonia lyase: Multiple helix dipoles implicated in catalysis. *Biochemistry* 43(36):11403–11416.

Cammack, R., Attwood, T.K., Campbell, P.N., Parish, J.H., Smith, A.D., Stirling, J.L., and Vella, F (2000) *Oxford Dictionary of Biochemistry and Molecular Biology*, revised edition. Oxford: Oxford University Press:98.

Castellarin SD, Di Gaspero G (2007) Transcriptional control of anthocyanin biosynthetic genes in extreme phenotypes for berry pigmentation of naturally occurring grapevines. *BMC Plant Biology* 7(1):1.

Chien S-C, Wu Y-C, Chen Z-W, Yang W-C (2015) Naturally occurring anthraquinones: Chemistry and therapeutic potential in autoimmune diabetes. *Evidence-Based Complementary and Alternative Medicine* (https://doi.org/10.1155/2015/357357).

Corder R, Douthwaite JA, Lees DM, Khan NQ, Viseu dos Santos AC, Wood EG, Carrier MJ (2001) Endothelin-1 synthesis reduced by red wine. *Nature* 414(6866):863–864.

Dixon RA, Paiva NL (1995) Stress-induced phenylpropanoid metabolism. *The Plant Cell* 7(7):1085.

Dregus M, Schmarr H-G, Takahisa E, Engel K-HJ (2003) Enantioselective analysis of methyl-branched alcohols and acids in Rhubarb (*Rheum rhabarbarum* L.) stalks. *Journal of Agricultural and Food Chemistry* 51(24):7086–7091.

Ferrer J-L, Jez JM, Bowman ME, Dixon RA, Noel JP (1999) Structure of chalcone synthase and the molecular basis of plant polyketide biosynthesis. *Nature Structural and Molecular Biology* 6(8):775.

Foust CM (2014) *Rhubarb: The Wondrous Drug*, Voloume 191. Princeton, NJ: Princeton University Press.

Fraser CM, Chapple C (2011) The phenylpropanoid pathway in Arabidopsis. *Arabidopsis Book* 9:20–211.

Han Y-S, van der Heijden R, Lefeber AW, Erkelens C, Verpoorte R (2002) Biosynthesis of anthraquinones in cell cultures of Cinchona 'Robusta' proceeds via the methylerythritol 4-phosphate pathway. *Phytochemistry* 59(1):45–55.

Han Y-S, Van der Heijden R, Verpoorte R (2001) Biosynthesis of anthraquinones in cell cultures of the Rubiaceae. *Plant Cell, Tissue and Organ Culture* 67(3):201–220.

Inoue K, Shiobara Y, Nayeshiro H, Inouye H, Wilson G, Zenk MH (1984) Biosynthesis of anthraquinones and related compounds in Galium mollugo cell suspension cultures. *Phytochemistry* 23(2):307–311.

Iwashina T, Omori Y, Kitajima J, Akiyama S, Suzuki T, Ohba H (2004) Ohba HJJoPR Flavonoids in translucent bracts of the Himalayan Rheum nobile (Polygonaceae) as ultraviolet shields. *Journal of Plant Research* 117(2):101–107.

Jin W, Wang YF, Ge RL, Shi HM, Jia CQ, Tu PFJR (2007) Simultaneous analysis of multiple bioactive constituents in Rheum tanguticum Maxim. ex Balf. by high-performance liquid chromatography coupled to tandem mass spectrometry. *Rapid Communications in Mass Spectrometry: An International Journal Devoted to the Rapid Dissemination of Up-to-the-Minute Research in Mass Spectrometry* 21(14):2351–2360.

Kemper, KJ (1999) Rhubarb root (*Rheum officinale* or *R. palmatum*). *The Center for Holistic Pediatric Education and Research. Longwood Herbal Task Force.* Available from: http://www. mcp. edu/herbal/default. htm

Kosikowska U, Smolarz H, Malm AJOLS (2010) Antimicrobial activity and total content of polyphenols of Rheum L. species growing in Poland. *Open Life Sciences* 5(6):814–820.

Krafczyk N, Kötke M, Lehnert N, Glomb MA (2008) Phenolic Composition of Rhubarb. *European Food Research and Technology* 228(2):187.

Phytochemistry 73

Leistner E (1974) [Isolation, identification and biosynthesis of anthraquinones in cell suspension cultures of Morinda citrifolia (author's transl)]. *Planta Medica* 1:214–224.

Leistner E (1981) Biosynthesis of plant quinones. *The Biochemistry of Plants* 7:403–423.

Leistner E (1995) Morinda species: Biosynthesis of quinones in cell cultures. In: *Medicinal and Aromatic Plants*, Volume VIII. Berlin/Heidelberg, Germany: Springer:296–307.

Lin Y-L, Wu C-F, Huang Y-T (2008) Phenols from the roots of Rheum palmatum attenuate chemotaxis in rat hepatic stellate cells. *Planta Medica* 74(10):1246–1252.

MacDonald MJ, D'Cunha GB (2007) A modern view of phenylalanine ammonia lyase. *Biochemistry and Cell Biology* 85(3):273–282.

Maeda H, Dudareva N (2012) The shikimate pathway and aromatic amino acid biosynthesis in plants. *Annual Review of Plant Biology* 63:73–105.

Malik S, Sharma N, Sharma UK, Singh NP, Bhushan S, Sharma M, Sinha AK, Ahuja PS (2010) Qualitative and quantitative analysis of anthraquinone derivatives in rhizomes of tissue culture-raised *Rheum emodi* Wall. plants. *Journal of Plant Physiology* 167(9):749–756.

Matsuda H, Morikawa T, Toguchida I, Park J-Y, Harima S, Yoshikawa MJB (2001) Antioxidant constituents from rhubarb: Structural requirements of stilbenes for the activity and structures of two new anthraquinone glucosides. *Bioorganic & Medicinal Chemistry* 9(1):41–50.

Munro AW, Girvan HM, McLean KJ (2007) Variations on a (t) heme—Novel mechanisms, redox partners and catalytic functions in the cytochrome P450 superfamily. *Natural Product Reports* 24(3):585–609.

Nikolau BJ, Ohlrogge JB, Wurtele ES (2003) Plant biotin-containing carboxylases. *Archives of Biochemistry and Biophysics* 414(2):211–222.

Pandey A, Misra P, Choudhary D, Yadav R, Goel R, Bhambhani S, Sanyal I, Trivedi R, Trivedi PK (2015) AtMYB12 expression in tomato leads to large scale differential modulation in transcriptome and flavonoid content in leaf and fruit tissues. *Scientific Reports* 5: 1–14.

Pandith SA, Dar RA, Lattoo SK, Shah MA, Reshi ZA (2018) Rheum australe, an endangered high-value medicinal herb of North Western Himalayas: A review of its botany, ethnomedical uses, phytochemistry and pharmacology. *Phytochemistry Reviews: Proceedings of the Phytochemical Society of Europe* 17(3):573–609.

Pandith SA, Dhar N, Rana S, Bhat WW, Kushwaha M, Gupta AP, Shah MA, Vishwakarma R, Lattoo SK (2016) Functional promiscuity of two divergent paralogs of type III plant polyketide synthases. *Plant Physiology* 171(4):2599–2619.

Pandith SA, Hussain A, Bhat WW, Dhar N, Qazi AK, Rana S, Razdan S, Wani TA, Shah MA, Bedi Y, Hamid A, Lattoo SK (2014) Evaluation of anthraquinones from Himalayan rhubarb (Rheum emodi Wall. ex Meissn.) as antiproliferative agents.) as antiproliferative agents. *South African Journal of Botany* 95:1–8.

Piotrowska H, Kucinska M, Murias MJMR (2012) Biological activity of piceatannol: Leaving the shadow of resveratrol. *Mutation Research/Reviews in Mutation Research* 750(1):60–82.

Püssa T, Raudsepp P, Kuzina K (2009) Polyphenolic Composition of Roots and Petioles of *Rheum Rhaponticum* L. *Phytochemical Analysis* 20(2):98–103.

Ragasa CY, Bacar JNB, Querido MMR, Tan MCS, Oyong GG, Brkljača R, Urban SJI (2017) JoP, research P. *Chemical Constituents of Rheum Ribes* L 9(1):65–69

Rehman H, Begum W, Anjum F (2014) Rheum emodi (Rhubarb): A fascinating herb. *Journal of Pharmacognosy and Phytochemistry* 3(2):89–94.

Rokaya MB, Maršík P (2012a) Active constituents in Rheum acuminatum and Rheum australe (Polygonaceae) roots: A variation between cultivated and naturally growing plants. *Biochemical Systematics and Ecology* 41:83–90.

Rokaya MB, Münzbergová Z, Timsina B, Bhattarai KRJ (2012b) Rheum australe D. Don: A review of its botany, ethnobotany, phytochemistry and pharmacology. *Journal of Ethnopharmacology* 141(3):761–774.

Rosler J, Krekel F, Amrhein N, Schmid J (1997) Maize phenylalanine ammonia-lyase has tyrosine ammonia-lyase activity. *Plant Physiology* 113(1):175–179.

Roupe KA, Remsberg CM, Yáñez JA, Davies NMJ (2006) Pharmacometrics of stilbenes: Seguing towards the clinic. *Current Clinical Pharmacology* 1(1):81–101.

Samappito S, Page JE, Schmidt J, De-Eknamkul W, Kutchan TMJP (2003) Aromatic and pyrone polyketides synthesized by a stilbene synthase from Rheum tataricum. *Phytochemistry* 62(3):313–323.

Sasaki Y, Nagano Y (2004) Plant acetyl-CoA carboxylase: Structure, biosynthesis, regulation, and gene manipulation for plant breeding. *Bioscience, Biotechnology, and Biochemistry* 68(6):1175–1184.

Schmidt J, Himmelreich U, Adam GJP (1995) Brassinosteroids, sterols and lup-20 (29)-en-2α, 3β, 28-triol from Rheum rhabarbarum 40(2):527–531

Singh R, Chauhan SJC, Chauhan SM (2004) 9,10-anthraquinones and other biologically active compounds from the genus *Rubia*. *Chemistry and Biodiversity* 1(9):1241–1264.

Stafford HA (1990) *Flavonoid Metabolism*. Boca Raton, FL: CRC Press.

Tayade A, Dhar P, Ballabh B, Kumar R, Chaurasia O, Bhatt R, Srivastava RJPA (2012) Rheum webbianum ROYLE: A potential medicinal plant FROM trans-Himalayan cold deserts of LADAKH, India. *Plant Archives* 12(2):603–606.

Wang J-B, Qin Y, Kong W-J, Wang Z-W, Zeng L-N, Fang F, Jin C, Zhao Y-L (2011) Identification of the antidiarrhoeal components in official rhubarb using liquid chromatography–tandem mass spectrometry. *Food Chemistry* 129(4):1737–1743.

Ye M, Han J, Chen H, Zheng J, Guo D (2007a) Analysis of phenolic compounds in rhubarbs using liquid chromatography coupled with electrospray ionization mass spectrometry. *Journal of the American Society for Mass Spectrometry* 18(1):82–91.

Ye M, Han J, Chen H, Zheng J (2007b) Analysis of phenolic compounds in rhubarbs using liquid chromatography coupled with electrospray ionization mass spectrometry. *Journal of the American Society for Mass Spectrometry* 18(1):82–91.

Zargar BA, Masoodi MH, Ahmed B, Ganie SAJFC (2011) Phytoconstituents and therapeutic uses of Rheum emodi wall. ex Meissn. *Food Chemistry* 128(3):585–589.

5 Pharmacology

Rhubarb has an extensive and age-old use in different traditional systems of medicine, including Unani, homeopathic, Chinese, and Tibetan systems, for the treatment of various kinds of ailments, besides being used as laxative, diuretic, and tonic. The recurrently occurring key biologically active phytoconstituents are reportedly known for cytotoxic, antimicrobial, antioxidant, antifungal, antitumor, antidiabetic, antiproliferative, and immunoenhancing activities (Rokaya et al. 2012; Pandith et al. 2014; Foust 2014). A detailed report about pharmacological activities (target and reported mechanism of action) of the individual flavonoid and anthraquinone chemical constituents from *Rheum* is available in one of our recent comprehensively compiled review articles (Pandith et al. 2018)—and references therein. However, and as a supplement to the previous compilation, here we present a general account of different biological activities associated with rhubarb, its crude extracts, or the individual phytoconstituents, in a broader perspective.

5.1 ANTIMICROBIAL ACTIVITY

Plant natural products, from extracts to lead molecules as drugs, have played a significant role in combating the alarming situation of resistance developed by many microbial agents against various marketed drugs and antibiotics. A perusal of the literature shows a greater surge of the scientific community toward evaluating the effects of various plant-based, as well as microbial, extracts and products against a wide range of disease-causing agents, the recalcitrant microbial strains in particular. In rhubarb, anthraquinone derivatives are reportedly known as the major and effective antimicrobial agents (Pandith et al. 2018; Bilal et al. 2013). In fact, the anthraquinone constituents like rhein, physcion, chrysophanol, and aloe-emodin have proven effective against common pathogenic bacterial species, viz. methicillin-resistant *Staphylococcus aureus* (Alaadin et al. 2007) and *Escherichia coli* K12 (Hatano et al. 1999), etc. Aloe-emodin had also been shown to inhibit growth of *H. pylori* through NAT N-acetyltransferase enzyme inhibition (Sydiskis et al. 1991), whereas rhein is known to have an antibacterial effect against *Bacteroides fragilis* and different acid-fast and gram-positive bacteria (Jong-Chol et al. 1987; Lewis and Elvin-Lewis 1977). Besides anthraquinones, the potent phytoceutical flavonoids, such as myricetin, quercetin, and rutin are also known to possess antibacterial activity. Myricetin has been shown to strongly inhibit DNA and RNA synthesis in *Proteus vulgaris* and *S. aureus*, respectively (Mori et al. 1987; Cushnie and Lamb 2005). Quercetin, on the other hand, inhibits growth of *E. coli* through DNA gyrase topoisomerase II inhibition (Kaul et al. 1985) and *Bacillus subtilis* and *Rhodobacter sphaeroides* through bacterial membrane dissipation (Hilliard et al. 1995; Mirzoeva et al. 1997).

DOI: 10.1201/9780429340390-5

Ibrahim et al. have shown that the ethanolic and benzene extracts of *R. australe* exhibit significant activity against various microbial agents, *Helicobacter pylori* in particular, and both under *in vivo* and *in vitro* conditions (Ibrahim et al. 2006). Nonetheless, use of benzene is restrained by certain agencies, including the National Toxicology Program (NTP), International Agency for Research on Cancer (IARC), and US Environmental Protection Agency (EPA), as they regard the chemical has carcinogenic effects on humans. Crude methanolic and n-butanol extracts of *R. palmatum* were tested against seven bacterial genera. It was found that methanolic extract of *R. palmatum* does possess antibacterial properties against *E. coli, Pseudomonas aeuroginosa, Shigella dysenteriae, Klebsiella pneumonia, Bacillus subtilis, S. aureus,* and *Micrococcus roseus* with the minimum inhibitory concentrations (MIC) of 50 to 175 µg/ml (Aly and Gumgumjee 2011). Indeed, the antibacterial effects of rhubarb have been ascribed to their capacity to inhibit the electron transport system in bacteria through multiple enzyme inhibitions in mitochondria (Chen and Chen 1987). In a study, rhubarb tablets cured 66% of the adult patients (number, 157) suffering from gonorrhea; though, without reporting co-therapies and the diagnostic criteria (Chen et al. 1991). Extracts of *R. officinale* was found to have significant activity against *Bacteroides fragilis* (a major anaerobic microorganism in the human intestines), and the purified active substance was identified as rhein (Jong-Chol et al. 1987).

Food security exists when the whole of mankind has physical, social, and economic access to safe, sufficient, and nutritious food. However, this access has been limited by the negative impacts on human health and environment which come into being upon application of hazardous pesticides, particularly fungicides, in traditional agricultural production systems to curb various diseases of crops and vegetables. Nevertheless, medicinal and aromatic plants, chiefly plant extracts with antimicrobial properties, have fascinated attention in the field of pest management (Sales et al. 2016). Many studies have demonstrated the active role of various *R. australe* extracts and the isolated bioactive anthraquinones against different fungal species, including *Aspergillus fumigatus, A. niger, Candida albicans,* and *Rhizopus oryzae,* etc. (Agarwal et al. 2000; Babu et al. 2003). In another experiment on wool yarns dyed with the extracts of *R. australe,* it was found that antifungal activity of the extracts against *C. albicans* and *C. tropicalis* significantly increased with the increase in concentration of the dyes (Khan et al. 2012). Further, crude methanolic and n-butanol extracts of *R. palmatum* were tested against five fungal genera. It was found that methanolic extract of *R. palmatum* does possess antifungal properties against *C. albicans, C. tropicalis, Cryptococcus neoformans, Alternaria solani, Fusarium oxysporum,* and *A. niger* with the minimum inhibitory concentrations (MIC) of 50 to 175 µg/ml (Aly and Gumgumjee 2011).

Some of the recent investigations have highlighted the efficacy of anthraquinone derivatives isolated from the rhizomes of *R. australe* as important antibacterial (Lu et al. 2011) and antiviral (Hussain et al. 2015) agents. The extracts from rhubarb rhizomes have shown virucidal effect against HSV-I, polio, influenza, and the measles virus *in vitro* (Sydiskis et al. 1991; May and Willuhn 1978; Kurokawa et al. 1993; Wang et al. 1996). Indeed, *R. palmatum* extracts are known

Pharmacology 77

as a potential drug against SARS coronaviruses due to their inhibitory activity against 3CL protease (Miraj 2016). Among the anthraquinone derivatives, emodin is a novel and safe antiviral agent for human respiratory and entero-viruses. It possesses the antiviral activity against RSV (respiratory syncytial virus) and CVB5 (coxsackievirus B5) infections by decreasing the mRNA expression of tumor necrosis factor-α (IFN-α) and enhancing TNF-γ expression significantly (Liu et al. 2015). It also inhibits the replication in herpes simplex virus (HSV-1, HSV-2), and hepatitis B viruses possibly by inhibition of UL-12 and CK-2 activity essential for viral replication and assemblage and disrupting the lipid bilayer (Xiong et al. 2011; Saito et al. 2012). Another major anthraquinone constituent, aloe-emodin, has been shown to be very effective against enveloped viruses like HSV-1 and HSV-2, pseudo rabies virus, influenza virus, and varicella zoster virus possibly by disrupting their envelopes. It has an inhibitory effect against entero virus 71 and JEV (Japanese encephalitis virus) caused by interferon signaling (Jassim and Naji 2003; Lin et al. 2008). Further, chrysophanol has been shown to inhibit poliovirus types 2 and 3, and picornaviridae at an early stage in the viral replication cycle (Jassim and Naji 2003; Semple et al. 2001). A 0.03% solution of rhein has also been shown to possess molluscicidal against *Bulinus globosus*, *Oncomelania hupensis*, *Bulinus globosus*, and *Biomphalaria glabrata* (Liu et al. 1997). Moreover, the foremost flavonoid from rhubarb, quercetin, has also proven effective against respiratory syncytial virus (RSV), para-influenza virus-3 Pf-3, polio, herpes simplex virus, and HSV-1 viruses through reduction in intracellular replication of certain RNA (Plaper et al. 2003). Additionally, another phenolic agent, kaempferol, has also been found to be effective against Japanese encephalitis virus through inhibition of viral replication and protein expression (Zhang et al. 2012). In general, the above studies, primarily based on the effectiveness of crude rhubarb extracts, possibly show the synergistic and multitude action of these extracts against various pathogens and could likely pave the way for the use of crude extracts on various crops. Nevertheless, investigating the mechanism of action of these extracts and/or their individual constituents may help in generating better, novel, and green antimicrobial agents.

5.2 ANTICANCER ACTIVITY

Clonal disorder cancer is a pathological condition identified by the World Health Organization (WHO) as one of the major threats to human health and development. In the United States alone, one in four deaths is due to cancer. However, phytoconstituents have found a use as efficient chemotherapeutic and/or chemopreventive agents for the effective treatment of a vast diversity of cancers. Indeed, phytomedicine offers a great deal to the health system as presently about 25% of the pharmaceutical prescriptions in the United States contain at least one plant-derived ingredient (Pandith et al. 2014). The methanolic and aqueous extracts of the rhizomes of *R. australe* were evaluated *in vitro* for their cytotoxic potential against breast (MDA-MB-435S), liver (Hep3B), and human prostate (PC-3) cancer cell lines to check their efficacy in combating tumors. The extracts, methanolic extract in particular, exhibited concentration-dependent activity (Rajkumar et al. 2011a) and also showed the ability to induce and push the cells toward apoptosis (Rajkumar et al. 2011b). In several such *in vitro* studies utilizing various human cancer cell lines, besides the normal cell lines as control, crude extracts of *R. australe* have been shown to display anti-metastatic potential (Kumar et al. 2012), cancer-cell-specific toxicity (Kumar et al. 2015; Kumar et al. 2013), and anti-angiogenic effects (Hsu and Chung 2012). In another study, besides various extracts, the major anthraquinone constituents, viz. maesopsin, naringenin chalcone, and naringenin from *R. australe*, were shown to demonstrate anti-angiogenic potential in zebra fish extracts (He et al. 2009). In another study, major bioactive anthraquinone constituents are known as effective apoptotic agents in various human and animal cancer cell lines (Lu et al. 2010). A perusal of the literature shows that among the key anthraquinone constituents, the majority of anti-tumor studies have been focused on maesopsin which has been demonstrated to display promising results in both *in vivo* and *in vitro* assay-based investigations. It was reported to induce apoptosis in cervical HeLa cells by increasing p53 levels (Füllbeck et al. 2005). Earlier studies have shown that the compound selectively inhibits casein kinase II (CKII), a Ser/Thr kinase by competitively binding to the ATP-binding pocket of the kinase against ATP (Yim et al. 1999). It is also known to constrain cell invasiveness through activator protein-1 (AP-1) and nuclear factor (NF)-κB signaling pathway suppression (Huang et al. 2004). Furthermore, in one of our own investigations focused on methanolic extracts of rhizomes, maesopsin, and rhapontigenin displayed significant antiproliferative activity against various human cancer cell lines, probably by reducing cell viability and rising mitochondrial membrane potential ($\Delta\psi$m) loss (Pandith et al. 2014).

Emodin, a natural and integral component of rhubarb, suppresses prostate cancer cell growth by targeting the androgen receptor. Degradation of the androgen receptor is induced in a ligand-dependent manner through a proteasome-mediated pathway (Cha et al. 2005). The compound induces apoptosis and inhibits growth in HepG2/C3A, PLC/PRF/5, and SKHEP-1 human hepatoma cell lines through activation of caspase-3, p53, p21, and Fas/APO-1 systems (Zhang et al. 1998), and human cervical cancer cell lines HeLa, CaSki, ME-180, and Bu25TK through ribose polymerase cleavage and activation of caspase-9 (Cha et al. 2005). It also represses metastasis-associated properties in human breast cancer cell lines MDA-MB453 and MCF-7 (Srinivas et al. 2003, 2007).

Aloe-emodin is an active component of the rhizomes of rhubarb species. It induces apoptosis in different human cancer cell lines. It induces apoptosis and DNA fragmentation in lung squamous cell carcinoma cell line CH27 (Lee et al. 2001). Apoptosis (programmed cell death) is characterized by expression and translocation of Bcl-2 family proteins, release of cytochrome c from mitochondrial inter-membrane space followed by activation of different caspases. One of the caspase cascades is initiated by caspase 8 and 9 resulting in activation of effecter caspases-3, 6, and 7 and the second cascade by Bcl-2 family proteins. The compound also blocks angiogenesis in colon adenocarcinoma cells by down-regulating Rho, Band, VEGF, and MMP-2/9 via reduced DNA binding activity of NF-jB—WiDr

Pharmacology

(Kuo et al. 2002). It has been shown to induce cell apoptosis, arrest the cell cycle at the G2/M phase and cell proliferation inhibition in human leukemia HL-60 cell line. The aloe-emodin treatments in these cells causes an increase in the levels of p27 and caspase-3 in addition to the DNA fragmentation (Chen et al. 2004). Teresa Pecere has reported induction of apoptosis, and *in vitro* and *in vivo* anti-neuroectodermal tumor activity by aloe-emodin through an energy dependent pathway (Pecere et al. 2000). It is also known to induce apoptosis in human nasopharyngeal carcinoma (NPC-TW076 and NPC-TW039 cancer cell lines), lung cancer (CH27 cancer cell line), urinary bladder cancer (T24 cell line), and human liver cancer (HepG2, and Hep3B cancer cell lines) through a caspase-3, 8, and 9-mediated activation system and p21, p53-dependent apoptotic pathway (Chen et al. 2004; Lin et al. 2006, 2010; Suboj et al. 2012; Kuo et al. 2002).

Another major anthraquinone, rhein, has been shown to induce apoptosis in human cervical cancer cell lines (Ca Ski) by encouraging decrease in Bcl-2 levels and increases in P53, Fas, p21, and Bax levels. It induces apoptosis in a caspase-dependent manner by up-regulating the activity of caspase-8 and 9 leading to the activation of caspase-3 which thereby causes DNA fragmentation (Ip et al. 2007). The rhein in combination with adriamycin inhibits ferricyanide and ferricyanide-induced proton release in human glioma cells in the brain and spine in a dose-dependent manner, therefore improving the therapeutic index of adriamycin, besides lowering its toxicity (Fanciulli et al. 1992). It also induces apoptosis and DNA damage followed by the inhibition of matrix metalloproteinase-9 and DNA repair-associated gene expression in the human tongue cancer cell line (SCC-4). It has shown to induce apoptosis effectively in human breast cancer (MCF-7 cell line) and in liver cancer (Hep G2 cell line). Some of the recent studies have proposed that the mechanism of action of rhein involves FOXO3a-mediated up-regulation of the Bim gene (Chen et al. 2004; Wang et al. 2015). Nevertheless, further mechanistic studies are required to validate the anticipated mechanism of action of this (and other) key anthraquinone constituent.

Besides anthraquinones, the chief flavonoid chemical constituents from rhubarb, viz. quercetin, kaempferol, daidzein, and myricetin, are reportedly known to possess anticancer activity that has been ascribed to their ability to inhibit cell cycle progression and down-regulate different anti-apoptotic and cancer regulatory compounds. Kaempferol and daidzein have been shown to inhibit cell cycle progression in "B"—human OCM-1 melanoma cells (Casagrande and Darbon 2001; Bandele and Osheroff 2007; Hashemi et al. 2012; Ravishankar et al. 2013; Constantinou et al. 1995). Quercetin, on the other hand, has been demonstrated to promote cancer cell apoptosis by down-regulating the heat shock protein (Hsp) 90. Decreased levels of Hsp90 induce cell death in prostate cancerous cells with decreased cell viability and activation of caspases in cancer cells with no effect on normal prostrate cells. The depletion of Hsp90 has also been shown to be due to short interfering RNAs similar to quercetin (Aalinkeel et al. 2008). It induces a G1 cell cycle block in "B"—human OCM-1 melanoma cells (Boege et al. 1996), inhibits topoisomerase I-catalyzed DNA relegation in human leukemia HL60 cell line (Cantero et al. 2006), and acts as a topoisomerase-II inhibitor in Chinese

80 Genus *Rheum*: A Global Perspective

hamster ovary AA8 cells and K562 cancer cell lines (Lu et al. 2006). The compound effectively inhibits rhodamine-123 efflux to revert back the multi-drug resistance (MDR) in MCF-7 breast cancer cell line (López-Lázaro et al. 2010). It is also a potent photosensitizer against UVB-induced skin cancer—murine melanoma cell line-B16F10 (Rafiq et al. 2015). All these studies suggest flavonoids as promising anticancer agents, besides being generally regarded as potent phytoceutical agents.

Other than anthraquinones and flavonoids, different stilbenoid compounds (stilbenes) from rhubarb, such as resveratrol, piceatannol, and rhapontigenin have also been shown to possess anticancer properties in various studies. Resveratrol is reported to induce cell death in varied cancer cell lines; it induces apoptosis in the HL60 cancer cell line (Dörrie et al. 2001; Kang et al. 2003), adult T-cells (Hayashibara et al. 2002), monocytic leukemia THP-1 cell line (Roupe et al. 2006; Tsan et al. 2000), breast cancer MDA-MB-231 cell line (Mgbonyebi et al. 1998; Scarlatti et al. 2008), breast cancer MCF-7 cell line (Mgbonyebi et al. 1998; Serrero and Lu 2001), colon cancer HCT-116 cell line (Mahyar-Roemer et al. 2002; Wolter et al. 2002), colon cancer Caco-2 cell line (Wolter et al. 2002), prostate DU-145 cancer cell line (Lin et al. 2002; Kampa et al. 2000), prostate cancer PC-3 cell line (Stewart and O'Brian 2004), melanoma A431 cell line (Ahmad et al. 2001; Adhami et al. 2001), melanoma-A375 cell line (Niles et al. 2003), melanoma SK-Mel-28 cell line (Niles et al. 2003; Larrosa et al. 2004), liver cancer HepG2 cell line (De Ledinghen et al. 2001), and ovarian cancer PA-1 cell line (Yang et al. 2003a) in a concentration-dependent manner. Piceatannol, another very important stilbene compound, has also been shown to possess anticancer properties. It is known to induce apoptosis in a human prostate cancer cell line (NRP-154) (Barton et al. 2004), inhibit tumor growth and metastasis in Lewis lung carcinoma (LLC)-bearing mice in a concentration of 10 µM (2.4 µg/mL) (Kimura et al. 2000), and for inhibition of preneoplastic lesions in a mouse mammary gland model with maximum inhibition at a concentration of 40.9 µM (10 µg/mL) (Waffo-Téguo et al. 2001). It plays a significant role in the inhibition of proliferation in the human colon adenocarcinoma cell (Caco-2) and human colorectal adenoma cell (HCT-116) in a concentration-dependent manner 0–200 µM (0–50 µg/mL) (Wolter et al. 2002).The stilbene rhapontigenin has been shown to possess an inhibitory effect on the human cytochrome P450 1A1 enzyme involved in biotransformation of different immuno-toxic and carcinogenic compounds at IC_{50} value of 0.4 µM (0.1 µg/mL) (Chun et al. 2001). It is also known as a potent anticancer compound in HepG2 liver cancer cells with an IC_{50} value of 80–120 µg/ml (Roupe et al. 2005). These investigations highlight the role of stilbenoids as possible anticancer agents, though further robust mechanistic studies, preferably based on *in vivo* models, need to be done to validate the above statements.

5.3 ANTIDIABETIC ACTIVITY

Diabetic patients experience an escalation of the transporters, viz. intestinal α-glucosidase and glucose transporter-2, that lead to a quick breakdown of

Pharmacology

disaccharides and thus glucose absorption by the body. The process may however enhance the levels of sugar in the blood to an abnormal level, paving the way for possible health complications in diabetics. Therefore, α-glucosidase inhibitors seem to play a crucial role in containing a good glycemic level within the body. To control this diabetic threat at the global level, development of innovative and operational therapy selections is needed that might even substitute for the standing therapeutic remedy consisting of an insulin sensitizer with α-glucosidase inhibitor. Currently, acarbose is such an inhibitor with an ordinary use, though a competitive one required in large quantities with reported intestinal instabilities and distress. This issue makes us to look for an alternate option to curb this global menace. In one of the recent investigations, Arvindekar et al. (2015) described the anti-hyperglycemic activity and α-glucosidase inhibitory effects of five foremost anthraquinone constituents isolated from *R. australe*, with aloe-emodin displaying the maximum action to lower the blood glucose levels. In another study of 120 patients with type 2 diabetes, *R. ribes* root was found to cause 39.63% reduction in blood glucose levels individually at the dose level of 350 mg when taken three times daily. The effect of the drug was analyzed after 12 weeks of drug administration. Further, the administration of *R. ribes* roots along with glibenclamide (a second-generation sulfonylurea antidiabetic agent with a potent and prolonged hypoglycemic effect used as an adjunct to the diet to lower the blood glucose in patients with non-insulin-dependent diabetes mellitus (type 2 diabetes)) has shown a reduction of 48.91% in blood glucose levels (Naqishbandi and Adham 2015). Moreover, hydroalcoholic extract of *R. turkestanicum* was shown to inhibit the development of myocardial damage, nephropathy, and liver injury damage in diabetics by obstructing lipid peroxidation, lowering the serum levels of lipids and glucose in a dose-dependent manner from 100–300 mg/kg in streptozotocin (a naturally occurring alkylating antineoplastic agent that is particularly toxic to the insulin-producing beta cells of the pancreas in mammals) induced diabetic rats (Hosseini et al. 2017).

1, 8-dihydroxyanthraquinones like rhein, emodin, aloe-emodin, chrysophanol, and physcion possess good anti-hyperglycemic activity. In treatments of auto immune diabetes (AID), T-cells and other leukocyte chemokine receptors (CXCR4, etc.) play an important role. Emodin and physcion have been shown to exhibit reduction in the migration of jurkat cells (an immortalized line of human T lymphocyte cells) in a human T-cell line mediated by CXCR4 in AID prophylaxis. The reduction is caused through inhibition of MEK1/2, MAPKs, MAPKK, and ERK ½ kinases in a dose dependent manner from 4 to 40 mg/kg (Chien et al. 2015). In another study, rhein and emodin have been shown to reduce C-C chemokine receptor type 4 (CXCR4)-mediated migration of jurkat cells at the half maximal inhibitory concentrations (IC_{50}) of 0.3 and 2.7 µg/mL, respectively. Rhein and emodin has also been shown to reduce C-C chemokine receptor type 5 (CCR5)-mediated migration of jurkat cells at IC_{50} of 0.75 and 2.5 µg/mL, respectively (Shen et al. 2012). Rhein has been widely used in the treatment of diabetes in animal and clinical experiments as well (Yang and Li 1993). Rhein has also proven effective in experimental diabetic nephropathy. It significantly decreased cellular hypertrophy, fibronectin synthesis, UDP-N-acetylglucosamine (UDP-GlcNAc) level, much higher glutamine: fructose 6-phosphate aminotransferase (GFAT) activity, and increased p21 and TGF-b1 (transforming growth factor-β1) expression in MCGT1 (transgenic mesangial cell line) cells cultured in normal glucose concentration. Moreover, rhein inhibits cellular hypertrophy and exerts its therapeutic role through decreased TGF-b1 and p21 expression and inhibition of the hexosamine pathway (Zheng et al. 2008).

When rhaponticin was administrated orally (125 mg/kg), oral glucose tolerance and highly reduced blood glucose levels in KK/Ay diabetic mice were recorded. It was seen that low density lipoprotein, insulin, raised plasma triglyceride, cholesterol, and non-esterified free fatty acid levels were also markedly attenuated. The study indicated that rhaponticin has huge antidiabetic effect and potentially could be used as a drug candidate to treat diabetes mellitus and related complications (Chen et al. 2009). Chrysophanol-8-O-β-D-glucopyranoside has been shown to activate glucose transport in insulin-stimulated myotubes (formed by the fusion of myoblasts) at a concentration up to 25 μM in a dose-dependent manner. Following its treatment, it was observed that Glut4 mRNA levels remain unchanged and tyrosine phosphorylation of insulin receptor increases due to inhibitory activity of tyrosine phosphatase 1B (a novel therapeutic target for type 2 diabetes mellitus, obesity, and related states of insulin resistance) at an IC_{50} value of 18.34 ± 0.29 μM. However, chrysophanol exerted mild glucose transport activity and phosphorylation of insulin receptor with increased Glut4 mRNA expression. So, both chrysophanol and chrysophanol-8-O-β-D-glucopyranoside were shown to have anti-diabetic properties (Lee and Sohn 2008).

The flavonoids such as epigallocatechin, epigallocatechin gallate, naringenin, and kaempferol are comparable to antidiabetic drugs used clinically. These flavonoids enhance insulin secretion and promote pancreatic B-cell proliferation, besides regulating glucose metabolism in hepatocytes, reducing inflammation and the oxidative stress. They are also known to decrease insulin resistance and increase glucose uptake in white adipose tissue and skeletal muscles through GLUT4 vesicle translocation to plasma membrane (Anand et al. 2010). Quercetin has been shown to reduce both glucose and glibenclamide-induced insulin secretion and phosphorylation of ERK1/2 at a concentration of 20 mmol/L using the INS-1 B-cell line. It also prevents B-cell dysfunction that is associated with glucose and glibenclamide-induced diabetes (Youl et al. 2010). Plant pigment (0.5%) has also been shown to improve the plasma insulin concentration and lower the streptozotocin-induced increase in blood glucose levels. It suppresses the expression of cyclin-dependent kinase inhibitor p21, WAF1/Cip1, and streptozotocin-induced expression of Cdkn1a in the pancreas. Its inhibition of Cdkn1a leads to improved pancreas and liver function, probably due to recovery of cell proliferation (Kobori et al. 2009). Another flavonoid, kaempferol, has a crypto-protective effect on human islets and beta cells. It reduces caspase-3 activity in human islets exposed to high glucose levels with maximum effect at 10 μM concentration. In islets and beta cells that are exposed to hyperglycemia, kaempferol improves the expression levels of Akt and Bcl-2 family anti-apoptotic molecules and restores

Pharmacology 83

high glucose-attenuated ATP and cAMP production. By protecting islets and pancreatic beta cell survival kaempferol may serve as a potent antidiabetic agent (Zhang and Liu 2011).

5.4 ANTI-INFLAMMATORY ACTIVITY

Inflammation, a common phenomenon, is the first reaction of living tissues to harmful stimuli such as pathogens, damaged cells, or irritants. It comprises systemic and local responses and is also critical for both innate and adaptive immunity. There is growing search for natural products, particularly phytoconstituents, interfering with these mechanisms by preventing or minimizing the prolonged effect of inflammation (Thangaraj 2016). Chauhan et al. (1992) studied the anti-inflammatory effect of petroleum ether, methanol, chloroform, and aqueous extracts of *R. australe* rhizomes to inhibit edema after a carrageenin injection in the left hind paw of Porton strain albino rats. An increase in the degree of inhibition was observed after carrageenin administration and was found to reach maximum at 5 h. The methanolic (500 mg/kg) and petroleum ether (500 mg/kg) extracts displayed an anti-inflammatory activity which was equivalent (59.6%) to the reference ibuprofen (50 mg/kg) at 5 h post carrageenin injection. Nevertheless, the aqueous (500 mg/kg) and chloroform (500 mg/kg) extracts were found to exhibit 21.3% and 45.3% edema inhibition respectively, at 5 h after carrageenin administration. This study had shown that methanolic and petroleum ether extracts at the effective dose of 500 mg/kg, p.o. can be useful in protecting rats against carrageenin-induced inflammation as effectively as the non-steroidal anti-inflammatory drug ibuprofen (50 mg/kg, p.o.). Nonetheless, due to some environmental/health issues related to the hexane, it is not generally considered a convenient solution to develop novel and efficacious drugs.

R. australe rhizome ethyl acetate extract was investigated for its role in the RAW 264.7 cell line for the expression of the nitric oxide synthase (NOS) gene and release of NO in murine macrophages. It was found that the extract up-regulates the expression of the NOS in RAW 264.7 and generates NO in a dose-dependent manner. It was further said that the ethyl acetate extract targets the transcription factor NF-κB for NOS induction and subsequent release of IL-12 and tumor necrosis factor-α (TNF-α) (Kounsar and Afzal 2010). In a similar experiment, crude extracts from four species of the genus *Rheum*, viz. *R. palmatum*, *R. nobile*, *R. officinale*, and *R. franzenbachii*, were used and it was found that *R. franzenbachii* possessed the strongest anti-inflammatory activity among the four tested rhubarb species (Cheon et al. 2009). Moreover, the anti-inflammatory activity of extracts from *R. palmatum* against different skin disorders was also studied. The extract in a dose-dependent manner inhibited the edema induced by 12-O-tetradecanoylphorbol-13-acetate (TPA), arachidonic acid and oxazolone-induced mouse ear edema in both single and multiple applications (Cuellar et al. 2001). In another study, an aqueous extract of *R. undulatum* L. was found to exhibit a distinct anti-inflammatory and vasorelaxant activity. It has shown to induce relaxation of the phenylephrine precontracted aorta with increased production of cGMP. In human umbilical vein endothelial cells (HUVECs), extract treatment attenuated TNF-α-induced NF-nB p65 translocation in a dose-dependent manner. It also suppressed expression of adhesion molecules like the intercellular adhesion molecule-1 (ICAM-1) and vascular cell adhesion molecule-1 (VCAM-1). Hence, the aqueous extract of *R. undulatum* was proposed to be a potent anti-inflammatory agent that suppresses the vascular inflammatory process via endothelium-dependent NO/cGMP signaling (Moon et al. 2006). Nonetheless, the precise mechanism of action of these extracts needs to be well understood which may possibly assist in the production of lead molecules acting as effective anti-inflammatory agents. In addition to the crude plant-based extracts, some of the major anthraquinone constituents from rhubarb have also been shown to display wide activity that pertains to the anti-inflammatory action. The study on the effect of aloe-emodin and rhein in TNF-α production and on single cell [Ca^{2+}] dynamics in lipopolysaccharide (LPS)-stimulated peritoneal macrophages and normal macrophages showed that rhein acts as a key in signal transduction pathways in LPS-stimulated macrophages by inhibiting the TNF-α production, and hence reduces the increase of [Ca^{2+}] in these cells. Whereas aloe-emodin at low concentrations inhibits the LPS-induced TNF-α production and attenuates the increase of [Ca^{2+}] induced by LPS, the effect decreases with increases in concentration (Yang et al. 2003b). Rhein, however, has been shown to enhance the inhibitory activity of aloe-emodin (Chen et al. 2007). Rhein-arginine (RA) has been shown to possess anti-inflammatory activity by significantly inhibiting hyperplasia, decrease in the levels of TNF-α and interleukin-1β (IL-1β), and by relieving adhesion of fibrous connective tissue in rats with ankylenteron (Yin et al. 1993). Because of its capacity to inhibit [Ca^{2+}] mobilization, aloe-emodin blocks the proliferation of T lymphocytes induced by IL-2 and phytohaemagglutinins (PHA) (LI et al. 2008). During acute pancreatitis, it also inhibits the inflammatory mediators to decrease the levels of IL-6, platelet-activating factor (PAF), and TNF-α in rats (Li et al. 2009). Emodin, an important anthraquinone in rhubarb, reduces inflammatory response in lungs by positively modulating the mRNA expression of IL-10 and negatively modulating the mRNA expression of IL-6 and IL-1β in the rat models (Zheng et al. 2013). RAW 264.7 cells, when stimulated by LPS in the presence of aloe-emodin, resulted in marked reduction of NO, IL-1β, and IL-6. Further, it was also showed that aloe-emodin down-regulates LPS-induced phosphorylation of ERK, JNK, Akt and p38, iNOS protein expression, and IκBα degradation. These studies suggest that aloe-emodin confers an anti-inflammatory effect in pro-inflammatory LPS-induced RAW 264.7 via inhibition of the PI3K, MAPK, and NF-κB pathways (Hu et al. 2014a).

5.5 ANTIOXIDATIVE ACTIVITY

Oxidative stress, an imbalance between pro-oxidants and antioxidants in the body, leads to enhanced production of free radicals, much above the detoxifying potential of local tissues. The resulting chain of destruction as a proximate effect of oxidative damage has been often linked to severe disease development and aging, prompting the scientific community to advocate the use of antioxidants to

Pharmacology 85

thwart or even reverse these conditions (Ristow 2014). Antioxidants are the compounds that are capable of either inhibiting or delaying the oxidation processes. They are used for the stabilization of foodstuffs, cosmetics, polymeric products, petrochemicals, and pharmaceuticals. Antioxidant compounds are involved in the defense mechanism of the organism, and these include enzymes, like catalase, superoxide dismutase, glutathione peroxidase, and non-enzymatic compounds, such as albumin, uric acid, bilirubin, and the highly conserved, cysteine-rich metal-binding proteins, viz. metallothioneins (Pisoschi and Negulescu 2011). For complete protection of the organism against reactive oxygen species (ROS), antioxidants taken as nutritional supplements or pharmaceutical products like vitamin E, β-carotene, vitamin C, flavonoids, and minerals play an important role (Litescu et al. 2011). The interest in antioxidants, particularly those intended to prevent the deleterious effects of free radicals and ROS in the human body, is increasing exponentially (Molyneux 2004). The health benefits of antioxidants in relation to cancer prophylaxis and therapy, radicals, and oxidative stress, have attracted considerable attention and longevity (Kalcher et al. 2009). However, natural antioxidants are biologically safe compared to synthetic antioxidants such as butylated hydroxytoluene (BHT) or butylated hydroxyanisole (BHA) which may cause DNA damage and prove toxic (Pandith et al. 2018).

The antioxidants from rhubarb, a good source of these phytoconstituents, are considered to be the good and promising candidates in contrast to the synthetic ones. Two Polish varieties of rhubarb, viz. Victoria and red malinowy, were accessed for their total antioxidant properties. Aqueous extracts of the two varieties were subjected to trolox equivalent antioxidant capacity (TEAC), 2,2'-azinobis-3-ethylbenzothiazoline-6-sulfonic acid (ABTS), 2,2-diphenyl1-picrylhydrazyl (DPPH), and ferric-reducing antioxidant power (FRAP) assays. The red malinowy rhubarb variety was characterized by the highest, and Victoria by the lowest, total antioxidant activity. The antioxidant property in these varieties was attributed to the percentage of flavonols (49.79–73.49 mg/100 g dm) in them (Kalisz et al. 2020). Various studies have been conducted to evaluate the antioxidant potential of *R. australe* and its bioactive phytoconstituents. A review of the related literature shows that different assays, including DPPH, inhibition on lipid peroxidation *in vitro*, hydroxyl radical scavenging activities, Fe^{3+}-reducing capacity, and hydrogen peroxide (H_2O_2), etc., have been used to examine the antioxidant potential of this plant species in the form of various crude extracts (Rajkumar et al. 2011a), as a constituent of herbal formulations (Ahmad et al. 2013) or of its individual constituents based on activity-guided assays directly (Hu et al. 2014b). All these studies have demonstrated the utility and efficiency of *R. australe* and its key constituents, including phenolics, anthraquinones, and stilbenoids, as potential antioxidative agents which have a good take for the food and pharmaceutical industries (Arvindekar and Laddha 2016; Mishra et al. 2014). Moreover, the methanolic and aqueous extracts of the rhizomes of *R. australe* of Nepalese origin were evaluated for their antioxidant efficacies. It was found that the aqueous extract contained less potent DPPH free radical scavenging activity than the methanolic extract (Gupta et al. 2014). Additionally, the antioxidant activity of the methanol and chloroform extract of roots and stems of *R. ribes* L. was evaluated following different antioxidant assays, namely DPPH radical scavenging, cupric reducing power (CUPRAC), FRAP, total antioxidant (lipid peroxidation inhibition activity), the b-carotene bleaching method, superoxide anion radical scavenging, and metal chelating activities. Methanol and chloroform extracts of roots exhibited higher activity with 84.1and 93.1% inhibitions while as stem extracts exhibited 82.0 and 82.2% inhibitions respectively (Öztürk et al. 2007). Moreover, ethyl acetate extracts of the same species were found a potential scavenger of DPPH radicals with an IC_{50} value of 10.92 µg/ml for root and 206.28 µg/ml for shoot (Uyar et al. 2014). Ethanolic extracts of root showed the highest antioxidant activity (IC_{50} = 4.73 ± 0.21 µg/mL) closer to the value of vitamin C (0.046 ± 0.05 µg/mL) compared to the aqueous extract (25.62 ± 0.85 µg/mL) (Abdulla et al. 2014).

The antioxidant activity (trolox equivalent antioxidant capacity) of the phenolic constituents in the roots of *R. officinale* was evaluated with tannins and gallic acid isolated as the predominant antioxidant phenolic constituents. It was found that the hydroxyl-anthraquinones with one hydroxyl group on the benzene ring of the anthraquinone structure have high radical scavenging effects, although their glycosylation was seen to reduce the scavenging activity (Cai et al. 2004). With *R. palmatum* decoction in rats, anthraquinones were subject to an extensive and rapid conjugation metabolism, and the serum metabolites of *R. palmatum* exhibited a potential antioxidant effect on 2,2'-azobis (2-amidinopropane hydrochloride)-induced hemolysis at pharmacologically relevant concentrations (Shia et al. 2009). Besides anthraquinones, the antioxidant activity of two stilbenoids, viz. piceatannol-4-O-β-D-glucopyranoside (PICG) and its aglycon piceatannol (PICE) isolated from *R. australe* rhizomes, was evaluated using different antioxidant evaluation assays. Both have shown very promising antioxidant effects with PICE bearing the highest activity which is ascribed to the presence of 3-hydroxyl groups (Chai et al. 2012). Moreover, another stilbenoid, rhapontigenin, from *R. undulatum* was evaluated for its capacity to scavenge the DPPH radical, intracellular ROS, and H_2O_2. It was found that rhapontigenin protected against cellular DNA damage and membrane lipid peroxidation induced by H_2O_2. The rhapontigenin inhibited the activity of activator protein 1 (AP-1) which is a redox-sensitive transcription factor, and increased phosphorylation of extracellular signal-regulated kinase (ERK) (Zhang et al. 2007).

A perusal of the literature shows some other investigations which were focused on the antioxidative effects of the anthraquinone constituents from rhubarb. The antioxidant potential of some of the major anthraquinones was evaluated based on the inhibition of peroxidation of linoleic acid, and the activity of these compounds followed the trend; aloe-emodin > rhein > emodin (Yen et al. 2000). The pretreatment of rhein (another anthraquinone compound) in the IEC-6 cells of rat models has shown that it suppresses apoptosis and caspase-3 activity and causes inhibition of cell viability and intracellular ROS induced by H_2O_2. It also enhances glutathione S-transferase activity, superoxide dismutase (SOD) activity, catalase activity, and the glutathione content. The inhibition of oxidative damage

Pharmacology 87

by rhein in IEC-6 cells is known to occur via PI3K/Akt and Nrf2/HO-1 pathways (Zhuang et al. 2019). Moreover, rhein has a strong oxidation-resisting characteristic. To evaluate it after intra-gastric administration in rat models with traumatic brain injury (TBI), rhein was identified in the brain tissue of the controlled cortical impact (CCI) rats. It elevated the SOD, catalase (CAT), glutathione (GSH) level, and GSH/GSSG ratio, and diminished the malondialdehyde (MDA), and glutathione disulfide (GSSG) levels, thereby inducing an antioxidative effect (Xu et al. 2017). Nevertheless, additional and advanced studies, both *in vitro* and *in vivo*, are a prerequisite to institute the mechanism of action (discrete or synergistic) of these herbal extracts and their constituents to recommend them as prospective antioxidative agents. Moreover, as some of the substitutions like position and number of hydroxyl groups have been allied with the antioxidant activities of phenolics, assessing these constituents for structure activity relationship (SAR) assays would complement the prerogatives of their antioxidant potential.

5.6 IMMUNOENHANCING ACTIVITY

Macrophages are known to exhibit different immuno-modulatory functions (tissue remodeling, removal of damaged cells and as defense agents against microbial invasions) through the release of ROS including nitric oxide (NO), TNF-α, and various cytokines like IL-10, IL-12, and IL-4, etc. Kounsar et al. (2011; Kounsar and Afzal 2010) studied the immuno-modulatory effects of ethyl acetate extracts of *R. australe* rhizomes through the release of various cytokines including the effects on signal transduction parameters which may provide a considerate way of regarding the treatment principles of the crude extract. The study revealed that crude extracts demonstrated a dose-dependent increase in the release of NO and cytokines (TNF- α, IL-12) and a decrease in the production of IL-10 in mouse macrophage (RAW 264.7) cell lines. Further, enhanced production of cytokines (TNF- α, 200 ng/ml; IL-12, 530 ng/ml) and down-regulation of IL-10 were shown to induce generation and proliferation of Th-1 cells while switching off the Th-2 immune system indicating the *R. australe* exhibit immuno-enhancing effect via Th-1 and Th-2 cytokine regulation *in vivo*. This study has good dividends for ethnobotany in the form of various medical formulations.

5.7 NEPHROPROTECTIVE ACTIVITY

The kidneys, the main target for heavy metals, play an important role in the elimination of xenobiotics, including heavy metals. They are usually later accumulated in the proximal tubule, often regarded as the main site of accumulation and injury ultimately leading to kidney damage. Even though it is not known whether metals such as arsenic (As), cadmium (Cd), and lead (Pb) serve any physiological role in the body, they have been recognized as effective nephrotoxic agents adversely affecting kidney function even at low levels. Renal toxicity stimulated by heavy metals is not gender-specific and can occur at any age. Sometimes, it can even lead to renal failure depending upon the dose and duration of exposure (Barbier et al. 2005; Conner and Fowler 1993). Additionally, a number of antibiotics including the tetracyclines, penicillins, and cephalosporins as well as aminoglycosides and sulphonamides are also known to act as potential nephrotoxins (Zargar et al. 2011).

Gentamicin is an aminoglycoside antibiotic with known nephrotoxic effects (Martinez-Salgado et al. 2007). Its consumption leads to the deficiency of antioxidant enzymes, tubular necrosis, reduction of renal blood flow, release of ROS and reactive nitrogen species, and induction of inflammatory pathways (Balakumar et al. 2010; Lopez-Novoa et al. 2011). It also increases malondialdehyde, urinary glucose and protein levels, urea and creatinine levels in the serum, and decreases in thiol in kidney tissue. It has been observed that in rats with 80 mg/kg/day gentamicin treatment, *R. turkestanicum* hydro-alcoholic extract at doses of 100 and 200 mg/kg body weight prevented the nephrotoxic effect of gentamicin through reduction in serum-creatinine/-urea levels, and malondialdehyde as well, besides increasing thiol levels and normalization of urine glucose and protein concentrations. In a similar study, hexachlorobutadiene was also used to induce nephrotoxicity. The study concludes that the extract works by decreasing the toxic metabolites of hexachlorobutadiene, increasing antioxidant activity or by inhibiting enzymes involved in the bioactivation of hexachlorobutadiene such as cysteine-S-conjugate β-lyase and glutathione-S-transferase (GST) (Boroushaki et al. 2019).

The nephroprotective activity of both water soluble (WS) and water insoluble (W-INS) fractions of methanolic extracts of *R. australe* rhizome has been determined against chemical-induced (CCl_4, gentamicin, mercuric chloride, and potassium dichromate) kidney damage in Wistar albino rats by monitoring the levels of urea nitrogen and creatinine in serum. WS extract had been shown to exhibit a significant nephroprotective effect on all segments (S1, S2, and S3) of the proximal tubule of the kidney against cadmium, mercury, and potassium dichromate-induced nephrotoxicity probably through antioxidant action of the tannins present in the fraction. However, W-INS was found to improve the nephrotoxicity induced by metals, viz. CCl_4 and mercuric chloride, while having a protective effect only on the S2 segment of proximal tubule in rat models. It can be inferred from the levels of creatinine, urea, and nitrogen in the serum that the nephrotoxicity could be reversed (Alam et al. 2005). Moreover, in an attempt to investigate the effects of ethyl acetate, petroleum ether, and n-butanol extracts of rhubarb on the urinary metabolite profile in rats with adenine-induced chronic kidney disease, it was found that these extracts reversed the abnormalities of urinary metabolite dysfunction in the pathway of purine, amino acid, choline, and taurine metabolisms. Further, it was also observed that the nephroprotective effect of ethyl acetate extract was stronger than other extracts. Certainly, the rhubarb extract improved histopathological abnormalities like inflammation and interstitial fibrosis with overall improvements in renal function (Zhang et al. 2015).

A Chinese herbal preparation WH30+ composed of *R. palmatum* in decoction with *Cordyceps sinensis* (Ascomycete fungus), *Salvia miltiorrhiza* (Lamiaceae), *Radix astragali* (Fabaceae), *Leonurus sibiricus* (Lamiaceae), *Epihedium macranthum* (Berberidaceae), and *Radix codonopsis pilosulae* (medicinal Korean herb)

Pharmacology

89

when administered to rats at the dose of 50 mg/kg/day for ten days was shown to reverse the effects of adenine-induced chronic renal failure and glycerol-induced acute renal failure in them (Ngai et al. 2005). In another study carried out to evaluate the clinical efficacy of rhubarb extracts in dogs with chronic renal failure, extract of *R. officinale* was administered to them under trade name Rubenal®. It was found that Rubenal® preparation induced significant improvements in serum phosphorus concentration, clinical score (by vet and clients), level of proteinuria, serum blood urea nitrogen, and creatinine levels, with no adverse effects in dogs after six months of treatment. *R. officinale* extract also improved the clinical signs of chronic renal failure, viz. hypertension, azotemia, hyperphosphoremia, and proteinuria, which offer its usage as a good supplementary drug for treating dogs with clinical and subclinical renal diseases (Kim and Hyun 2012). In a similar study based on the treatment of chronic renal failure (adenine-induced chronic tubule-interstitial nephropathy), three rhubarb extracts (ethyl acetate, petroleum ether, and n-butanol extracts of *R. officinale*) were used which improved renal injury and dysfunction, and also reversed plasma metabolite abnormalities. The ethyl acetate extract of rhubarb also attenuated up-regulation of pro-fibrotic proteins including α-SMA, TGF-β1, PAI-1, FN, CTGF, and collagen-1, with overall improvements in fatty acid, glycerophospholipid, and amino acid metabolisms (Zhang et al. 2016).

In addition to the extracts from rhubarb, some of its individual chemical constituents have also been shown to act as positive nephron-protection agents against various renal toxins/ailments. Emodin is one of the potent nephroprotective agents mainly because of its antioxidant and anticancer actions. At a dose level of 0.5 mM, it restored the cisplatin-induced glutathione depletion and total antioxidant capacity in cultured human kidney (HEK 293) cells. It also augments the cisplatin-induced inhibition of various enzymes which include glutathione peroxidase, catalase, glutathione reductase, glutathione S-transferase, and superoxide dismutase (Waly et al. 2013). Another anthraquinone, rhein, is also a potent nephroprotective agent. Rhein, an anti-inflammatory effect-causing bioactive component in Niuhuang Shangqing tablets (NHSQ)—a well-known over-the-counter traditional Chinese medical preparation/drug composed of 19 herbs with rhubarb as one of the major ingredients—was orally administered in male and female Sprague-Dawley rats (220–240 g) in addition to the NHSQ and rhubarb individually. Plasma samples were harvested at different time points for pharmacokinetic (PK) analysis, and samples from plasma, urine, and visceral organs (liver, heart, brain, and kidney) of rats were collected for metabolic profile analysis. Following oral administration of NHSQ, and in comparison to that of pure rhein and rhubarb, the pharmacokinetic properties of rhein showed significant variation. The NHSQ was found to be superior to the rhubarb or pure rhein in PKs due to the synergetic effects of rhubarb, peppermint, and chrysanthemum (components of NHSQ). Out of the five metabolic pathways, viz. reduction, glucuronidation, methylation, sulfation, and glucosidation, rhein metabolism was seen to chiefly involve glucuronidation, methylation, and sulfation, with significant increase in its C_{max} (the peak plasma concentration of a drug after administration) and area under the plasma concentration-time curve (AUC) after oral administration of NHSQ compared with that of rhubarb and pure rhein. While revealing the effects of complex ingredients in NHSQs on the pharmacokinetics of rhein, the study provides a significant basis to further develop NHSQ and guide for its possible clinical applications (Zhang et al. 2019a). In a related study by the same group (Zhang et al. 2019b), while employing ultra-high-performance liquid chromatography/quadrupole time-of-flight mass spectrometry (UHPLC-Q-TOF-MS/MS) in combination with the *in silico* approach, the identity of *in vivo*-absorbed flavonoids and metabolites profile in biological samples was assessed following oral administration of NHSQTs in Sprague–Dawley rats. In nutshell, these studies would help in understanding the mechanism of the action of NHSQ, besides revealing its *in vivo* effective components for further clinical utility. Moreover, in a study, nephropathy and hyperuricemia (excess of uric acid in the blood) was induced by adenine and ethambutol in mice. The administration of rhein in these mice significantly decreased the serum uric acid level by increasing the excretion of urinary uric acid and inhibiting the xanthine oxidase activity. Rhein also reduced the symptoms of nephropathy by inhibiting the expression of transforming growth factor-$\beta 1$, and reducing the release of pro-inflammatory cytokines, including prostaglandin E2, interleukin 1β, and tumor necrosis factor-α. This conferred that rhein is a potent nephroprotective and anti-hyper-uricemic agent for clinical applications. Rhein has also been shown to inhibit human organic anion transporters (hOAT)-1 and hOAT-3, with an IC$_{50}$ value of 77.1 $\pm$ 5.5, which indicates its possible role in detoxification and reversal of nephrotoxicity with full retrieval of renal functions associated with hOAT abnormalities in human kidneys (Meng et al. 2015).

5.8 HEPATOPROTECTIVE ACTIVITY

The liver, the largest vital organ in the human body, functions as the chief site for the metabolism of nutrients and the excretion of xenobiotics to maintain homeostasis in the body. However, it is constantly and variedly exposed to environmental toxins, including alcohol, and often abused by poor drug habits that lead to various liver ailments like cirrhosis, fibrosis, hepatitis, and even hepatic cancer. The degree of derangement of the liver by disease or hepatotoxins is usually measured by the level of alkaline phosphatase (ALP), glutamate pyruvate transaminase (ALT), glutamate oxaloacetate transaminase (AST), albumin, bilirubin, and whole liver homogenate (Zargar et al. 2011). Although the plant-based preparations are mainly employed for the treatment of various liver disorders; modern medicine seems diminutive against the rising incidence of hepatic ailments (Ahsan et al. 2009; Gnanadesigan et al. 2017).

The alcohol-based liver diseases are becoming an alarming global health problem (Shukla et al. 2013) that pave the way for strategies which aim to reduce hepatic inflammation and fat accumulation caused by alcohol consumption to block the evolution of this health hazard. Equally and interestingly, the significance of herbs rich in bioactive constituents as new therapeutic agents is exponentially increasing (Neyrinck et al. 2014). One such highly medicinal herb is rhubarb, which alone or in decoction with other unrefined drugs, has been used from ancient times in China for the treatment of cholestatic hepatitis (Hu 1986). It is known to protect against chronic liver injury including the development of liver fibrosis that leads to the development of hepatocellular liver cirrhosis and carcinoma. Rhubarb rhizome extracts significantly slow down liver fibrosis through the inhibition of stellate cells resulting in down-regulating the expression of type 1 pro-collagen mRNA, α-smooth muscle actin (α-SMA) and tissue inhibitors of metalloproteinases-1 and 2 (TIMP-1 and 2) (Jin et al. 2005; Wang et al. 2011).

Ibrahim et al. (2008) studied the hepatoprotective effects of ethanolic extracts of *R. australe* by inducing hepatotoxicity in primary cultures of rat hepatocytes by carbon tetrachloride (CCl_4). The same were then treated with increasing concentration (10, 50, and 100 µg/ml) of ethanolic extract in a time-dependent manner (6, 12, and 24 h). After the treatment, there was a noticeable upsurge in the release of lactate dehydrogenase (LDH) and glutamate pyruvate transaminase (GPT) in a concentration-dependent manner, thereby confirming hepatoprotective activity. The same ethanolic extract was also used to determine the hepatoprotective activity against CCl_4-induced liver damage in Wistar male adult rats. A considerable increase in the serum activities of ALP, ALT, and AST was observed with a similar increase in the total bilirubin. Significant reduction in the above elevated parameters was observed upon oral administration of *R. australe* (3.0 g/kg, p.o.) and the liver was seen to restore to its normal size and pattern. Later, in 2009, a study was carried out to assess the effect of aqueous extract of *R. australe* on hepatic microsomal enzyme P450 for the pentobarbital-induced sleeping time and the lethal effects of strychnine in four groups of rats wherein it was found that the extract of *R. australe* inhibits hepatic microsomal P450 enzyme and causes significant prolongation in pentobarbital-induced sleeping time conferring the hepatoprotective role of the extract (Tahir et al. 2009). Importantly, in another experiment by Ibrahim et al. (2012), *S. mukorossi* pericarp (2.5 mg/mL) and *R. australe* rhizomes (3.0 mg/mL) were shown to protect the rats from liver cirrhosis induced by CCl_4. The results were inferred from histopathological evidence and activities of the serum marker enzymes, viz. AST, ALP, and ALT. In another experiment, methanolic and chloroform extracts of *R. australe* rhizomes were subjected to *in vivo* hepatoprotective activity in male Wistar rats against paracetamol (acetaminophen)-induced toxicity. Silymarin (50 mg/kg BW) was used as a control hepatoprotective drug. The extracts retained the normal liver by effective reduction in the concentration AST, ALT, ALP, and bilirubin (Akhtar et al. 2016). In a similar experiment, aqueous and alcoholic extracts were used, and it was found that the former shows a significant hepatoprotective effect against the CCl_4-induced hepatic damage, though only if given before metabolic activation of the CCl_4 (Tahir et al. 2009).

Additionally, low-dose *R. palmatum* decoctions have been shown to attenuate nonalcoholic fatty liver disease, reduce expression of lipogenic genes, and excess fat accumulation in the liver thereby alleviating hepatic steatosis by stimulating acetyl-CoA carboxylase AMPK activities (Yang et al. 2016). In related and recent investigations, rhubarb extract (0.3%) has been shown to decrease the oxidative stress, liver tissue injury, inflammatory disorders, and hepatic lipid accumulation caused by acute alcohol administration in male C57Bl6J mice. The extract also changed the microbial ecosystem in favor of *Parabacteroides goldsteinii* and *Akkermansia muciniphila* as well as the increase in tissue weight, crypt depth, and the expression of antimicrobial peptides (Neyrinck et al. 2014, 2017).

Emodin is a significant hepatoprotective agent. The rhubarb emodin has been shown to possess restoring activity on cholestatic hepatitis by antagonizing pro-inflammatory mediators and cytokines, improving hepatic microcirculation, inhibiting oxidative damage, controlling neutrophil infiltration, and reducing impairment signals (Ding et al. 2008). When emodin (20, 30, and 40 mg/kg, p.o.) was administered in acetaminophen (2 g/kg, p.o.) treated rats, it reduced the toxicity of acetaminophen by protecting acetaminophen-induced alterations after 24 h of administration. Acetaminophen administration caused elevation of serum transaminases, dehydrogenase, alkaline phosphatase, lactate, and bilirubin and serum protein with a decrease in hemoglobin and blood sugar levels. It also altered alkaline phosphatase, glutathione, and adenosine triphosphatase levels with intense lipid peroxidation. Emodin at a median dose of 30 mg/kg brings the hepatic enzyme levels to normal and protects the liver against acetaminophen-induced toxicity (Bhadauria 2010). The hepatoprotective effect of total anthraquinone glycoside content from dried rhizomes of *R. australe* (200–400 mg/mL) was analyzed in CCl_4-treated rats. The anthraquinone glycoside treatment was shown to improve the removal of bromsulphalein rate from hepatic cells followed by increased cell viability. Further, the pretreatment in a dose-dependent manner also showed the restoration of antioxidant and serum enzyme levels (Srinivasarao et al. 2015). Additionally, ethanolic extract of trans-resveratrol (20 mg/kg per day) and hydroxy-stilbenes from the roots of *R. rhaponticum* administered intraperitoneally in mice with hyper ethanolic hepato-toxicity exert a protective effect against liver damage in them. Trans-resveratrol and hydroxy-stilbenes inhibit oxidation of the polyunsaturated fatty acids in the blood and thereby attenuate the oxidative stress (Raal et al. 2009; Xing et al. 2011). Similar work has been done in *R. officinale* with comparable results (Wang et al. 2009). However, the mechanism of action of these extracts (and their individual constituents) needs to be evaluated as well as ruling out the possible adverse (toxicity) effects.

5.9 MISCELLANEOUS ACTIVITIES

Rhubarb has many other pharmacological properties other than those discussed above. It helps in regulating cholesterol levels. The stalk fiber from *R. rhaponticum* is hypo-lipidemic (Cheema et al. 2003) due to its high capacity to bind with bile-acids controlling cholesterol levels through up-regulating a rate-limiting enzyme 7 α-hydroxylase (cyp7a) in cholesterol metabolism (Goel et al. 1999). During pregnancy in women, rhubarb shows a protective effect against high blood pressure and significantly reduces vascular endothelial cell damage which is effective in treating high blood pressure (Zhang et al. 1994; Wang

Pharmacology

and Song 1999). Also, during menstrual flow rhubarb helps to move stagnated blood and stimulates the uterus, relieves endometriosis correlated symptoms, pain, and cramps, possessing estrogenic relative potential (Kang et al. 2008). Lindleyin (Usui et al. 2002), rhapontigenin, and deoxyrhapontigenin, extracted from *R. rhaponticum*, reduced the severity and frequency of hot flushes in peri-menopausal women (Papke et al. 2008). Moreover, lindleyin and rhapontigenin specifically bind to estrogen receptors ERα and ERβ, and may act as selective estrogen receptor modulators as well (Cosman and Lindsay 1999). In traditional Chinese medicine, aqueous extract of rhubarb is known to prevent the development of atherosclerosis by down-regulating the vascular expression of adhesion molecules and pro-inflammatory molecules via regulation of the endothelin system and nitric oxide production (Iizuka et al. 2004; Liu et al. 2008). Rhubarb is also known to show a laxative effect on the gut and large intestine by stimulating colonic motility and augmenting cellular permeability in colonic mucosa through inhibition of Na^+/K^+ ATPase pump, and chloride channels (Leng-Peschlow 1986; Yamauchi et al. 1993). Further, it promotes excitatory activity in the duodenum and colon (Jin et al. 1989) mediated via the cholinergic N receptor, M receptor, and L-type calcium channels (Yu et al. 2005). Coming to the rhubarb phyto-constituents, rhaponticin (a stilbenoid glucosidic form of the aglycone rhaponti-genin) and rhapontigenin extracted from rhubarb are effective against amyloid beta-induced neurotoxicity which forms senile plaques in the brains of patients with Alzheimer's disease. These act as neuro-protective agents maintaining cell viability and mitochondrial functionality, probably by regulating expression of the bcl-2 gene family, and hence are potential agents for the management of Alzheimer's disease (Misiti et al. 2006). On the other hand, rhein, extracted from rhubarb, has been shown to attenuate arylamine N-acetyltransferase (NAT) activity in *H. pylori* collected from peptic ulcer patients in a dose-dependent manner (Chung et al. 1998).

In nutshell, rhubarb has seen very wide pharmacological utility across boundaries with no recorded side effects. The pharmacological efficacy of this wondrous drug has primarily been attributed to its major bioactive phytoconstituents, mainly anthraquinones and stilbenoids, some of which have been tested both under *in vitro* and *in vivo* conditions. Nonetheless, such reports of *in vitro/vivo* action are inadequate to establish a valid pharmacological formula. Also, most of the studies related to rhubarb species are based at the level of herbal extracts and not the individual compounds. Importantly, the synergistic study-based effects of this herbal drug cannot be ruled out, wherein its formulation is based on the mixture of different medicinal herbs, including one or more species of *Rheum* as well. In this context, what is required is the mechanistic of the action of individual chemical constituents from the crude extracts and/or herbal formulations, and of the compounds (individual) when administered in pure form. Additionally, their bioavailability, dosage/toxicity levels, and other attributes like pharmacokinetics and pharmacodynamics need to be assessed well. This would somehow lead us some way further toward developing a more promising health care system for the overall betterment of human society.

REFERENCES

Aalinkeel R, Bindukumar B, Reynolds JL, Sykes DE, Mahajan SD, Chadha KC, Schwartz SA (2008) The dietary bioflavonoid, quercetin, selectively induces apoptosis of prostate cancer cells by down-regulating the expression of heat shock protein 90. *The Prostate* 68(16):1773–1789.

Abdulla KK, Taha EM, Rahim SM (2014) Phenolic profile, antioxidant, and antibacterial effects of ethanol and aqueous extracts of Rheum ribes L. roots. *Der Pharmacia Lettre* 6(5):201–205.

Adhami VM, Afaq F, Ahmad N (2001) Involvement of the retinoblastoma (pRb)–E2F/DP pathway during antiproliferative effects of resveratrol in human epidermoid carcinoma (A431) cells. *Biochemical and Biophysical Research Communications* 288(3):579–585.

Agarwal S, Singh SS, Verma S, Kumar S (2000) Antifungal activity of anthraquinone derivatives from Rheum emodi. *Journal of Ethnopharmacology* 72(1–2):43–46.

Ahmad N, Adhami VM, Afaq F, Feyes DK, Mukhtar H (2001) Resveratrol causes WAF-1/p21-mediated G1-phase arrest of cell cycle and induction of apoptosis in human epidermoid carcinoma A431 cells. *Clinical Cancer Research : An Official Journal of the American Association for Cancer Research* 7(5):1466–1473.

Ahmad W, Zaidi SMA, Mujeeb M, Ansari SH, Ahmad S (2013) HPLC and HPTLC methods by design for quantitative characterization and in vitro anti-oxidant activity of polyherbal formulation containing Rheum emodi. *Journal of Chromatographic Science* 52(8):911–918.

Ahsan MR, Islam KM, Bulbul IJ, Musaddik MA, Haque E (2009) Hepatoprotective activity of methanol extract of some medicinal plants against carbon tetrachloride-induced hepatotoxicity in rats. *European Journal of Scientific Research* 37(2):302–310.

Akhtar MS, Habib A, Ali A, Bashir S (2016) Isolation, identification, and in vivoevaluation of flavonoid fractions of chloroform/methanol extracts of Rheum emodi roots for their hepatoprotective activity in Wistar rats. *International Journal of Nutrition, Pharmacology, Neurological Diseases* 6(1):28.

Alaadin AM, Al-Khateeb EH, Jäger AK (2007) Antibacterial activity of the Iraqi Rheum ribes. Root. *Pharmaceutical Biology* 45(9):688–690.

Alam MA, Javed K, Jafri M (2005) Effect of Rheum emodi (Revand Hindi) on renal functions in rats. *Journal of Ethnopharmacology* 96(1):121–125.

Aly MM, Gumgumjee NM (2011) Antimicrobial efficacy of Rheum palmatum, Curcuma longa and Alpinia officinarum extracts against some pathogenic microorganisms. *African Journal of Biotechnology* 10(56):12058–12063.

Anand S, Muthusamy V, Sujatha S, Sangeetha K, Raja RB, Sudhagar S, Devi NP, Lakshmi B (2010) Aloe emodin glycosides stimulates glucose transport and glycogen storage through PI3K dependent mechanism in L6 myotubes and inhibits adipocyte differentiation in 3T3L1 adipocytes. *FEBS Letters* 584(14):3170–3178.

Arvindekar A, Laddha K (2016) An efficient microwave-assisted extraction of anthraquinones from Rheum emodi: Optimisation using RSM, UV and HPLC analysis and antioxidant studies. *Industrial Crops and Products* 83:587–595.

Arvindekar A, More T, Payghan PV, Laddha K, Ghoshal N, Arvindekar A (2015) Evaluation of anti-diabetic and alpha glucosidase inhibitory action of anthraquinones from Rheum emodi. *Food and Function* 6(8):2693–2700.

Babu KS, Srinivas P, Praveen B, Kishore KH, Murty US, Rao JM (2003) Antimicrobial constituents from the rhizomes of Rheum emodi. *Phytochemistry* 62(2):203–207.

Balakumar P, Rohilla A, Thangathirupathi A (2010) Gentamicin-induced nephrotoxicity: Do we have a promising therapeutic approach to blunt it? *Pharmacological Research* 62(3):179–186.

Pharmacology

Bandele OJ, Osheroff N (2007) Bioflavonoids as poisons of human topoisomerase IIα and IIβ. *Biochemistry* 46(20):6097–6108.

Barbier O, Jacquillet G, Tauc M, Cougnon M, Poujeol P (2005) Effect of heavy metals on, and handling by, the kidney. *Nephron. Physiology* 99(4):105–110.

Barton BE, Karras JG, Murphy TF, Barton A, Huang HF (2004) Signal transducer and activator of transcription 3 (STAT3) activation in prostate cancer: Direct STAT3 inhibition induces apoptosis in prostate cancer lines. *Molecular Cancer Therapeutics* 3(1):11–20.

Bhadauria M (2010) Dose-dependent hepatoprotective effect of emodin against acetaminophen-induced acute damage in rats. *Experimental and Toxicologic Pathology : Official Journal of the Gesellschaft Fur Toxikologische Pathologie* 62(6):627–635.

Bilal S, Mir M, Parrah J, Tiwari B, Tripathi V, Singh P, Mehjabeenc AA (2013) Rhubarb: The wondrous drug. A review. *International Journal of Pharmacy and Biological Sciences* 3(3):228–233.

Boege F, Straub T, Kehr A, Boesenberg C, Christiansen K, Andersen A, Jakob F, Köhrle J (1996) Selected novel flavones inhibit the DNA binding or the DNA religation step of eukaryotic topoisomerase I. *Journal of Biological Chemistry* 271(4):2262–2270.

Boroushaki MT, Fanoudi S, Rajabian A, Boroumand S, Aghaee A, Hosseini A (2019) Evaluation of rheum turkestanicum in hexachlorobutadien-induced renal toxicity. *Drug Research* 69(08):434–438.

Cai Y, Sun M, Xing J, Corke H (2004) Antioxidant phenolic constituents in roots of Rheum officinale and Rubia cordifolia: Structure– radical scavenging activity relationships. *Journal of Agricultural and Food Chemistry* 52(26):7884–7890.

Cantero G, Campanella C, Mateos S, Cortés F (2006) Topoisomerase II inhibition and high yield of endoreduplication induced by the flavonoids luteolin and quercetin. *Mutagenesis* 21(5):321–325.

Casagrande F, Darbon J-M (2001) Effects of structurally related flavonoids on cell cycle progression of human melanoma cells: Regulation of cyclin-dependent kinases CDK2 and CDK1. *Biochemical Pharmacology* 61(10):1205–1215.

Cha T-L, Qiu L, Chen C-T, Wen Y, Hung M-C (2005) Emodin down-regulates androgen receptor and inhibits prostate cancer cell growth. *Cancer Research* 65(6):2287–2295.

Chai Y-Y, Wang F, Li Y-L, Liu K, Xu H (2012) Antioxidant activities of stilbenoids from Rheum emodi Wall. *Evidence-Based Complementary and Alternative Medicine* (https://doi.org/10.1155/2012/603678).

Chauhan N, Kaith BS, Mann S (1992) Anti-inflammatory activity of Rheum australe roots. *International Journal of Pharmacognosy* 30(2):93–96.

Cheema SK, Goel V, Basu TK, Agellon LB (2003) Dietary rhubarb (Rheum rhaponticum) stalk fibre does not lower plasma cholesterol levels in diabetic rats. *British Journal of Nutrition* 89(2):201–206.

Chen C, Chen Q (1987) Biochemical study of Chinese rhubarb. XIX. Localization of inhibition of anthraquinone derivatives on the mitochondrial respiratory chain. *Yao xue xue bao= Acta Pharmaceutica Sinica* 22(1):12.

Chen H, Hsieh W-T, Chang W, Chung J-G (2004) Aloe-emodin induced in vitro G2/M arrest of cell cycle in human promyelocytic leukemia HL-60 cells. *Food and Chemical Toxicology: An International Journal Published for the British Industrial Biological Research Association* 42(8):1251–1257.

Chen J, Ma M, Lu Y, Wang L, Wu C, Duan H (2009) Rhaponticin from rhubarb rhizomes alleviates liver steatosis and improves blood glucose and lipid profiles in KK/Ay diabetic mice. *Planta Medica* 75(5):472–477.

Chen L-J, Sun W-W, Hu F, WANG X-Y, LIU H-J, YANG W-X (2007) Effect of aloe-emodin on $[Ca^{2+}]$ I and TNF-alpha production in peritoneal macrophages of rats. *Chinese Traditional and Herbal Drugs* 38(9):1359.

Chen Q, Han G, Zhuang X, Huang B (1991) Clinical observation on 157 subjects of gonorrhea treated with rhubarb preparation. *Journal of China Pharmaceutical University* 22:292–294.

Cheon M-S, Yoon T-S, Choi G-Y, Kim S-J, Lee A, Moon B-C, Choo B-K, Kim H-K (2009) Comparative study of extracts from rhubarb on inflammatory activity in RAW 264.7 cells. *Korean Journal of Medicinal Crop Science* 17(2):109–114.

Chien S-C, Wu Y-C, Chen Z-W, Yang W-C (2015) Naturally occurring anthraquinones: Chemistry and therapeutic potential in autoimmune diabetes. *Evidence-Based Complementary and Alternative Medicine* (https://doi.org/10.1155/2015/357357).

Chun YJ, Ryu SY, Jeong TC, Kim MY (2001) Mechanism-based inhibition of human cytochrome P450 1A1 by rhapontigenin. *Drug Metabolism and Disposition: The Biological Fate of Chemicals* 29(4):389–393

Chung JG, Tsou MF, Wang HH, Lo HH, Hsieh SE, Yen YS, Wu LT, Chang SH, Ho CC, Hung CF (1998) Rhein affects arylamine N-acetyltransferase activity in Helicobacter pylori from peptic ulcer patients. *Journal of Applied Toxicology: An International Forum Devoted to Research and Methods Emphasizing Direct Clinical* 18(2):117–123.

Conner E, Fowler B (1993) Mechanisms of metal-induced nephrotoxicity. *Toxicology of the Kidney* 1:437–457.

Constantinou A, Mehta R, Runyan C, Rao K, Vaughan A, Moon R (1995) Flavonoids as DNA topoisomerase antagonists and poisons: Structure-activity relationships. *Journal of Natural Products* 58(2):217–225.

Cosman F, Lindsay R (1999) Selective estrogen receptor modulators: Clinical spectrum. *Endocrine Reviews* 20(3):418–434.

Cuellar M, Giner R, Recio M, Manez S, Rıos J (2001) Topical anti-inflammatory activity of some Asian medicinal plants used in dermatological disorders. *Fitoterapia* 72(3):221–229.

Cushnie TT, Lamb AJ (2005) Antimicrobial activity of flavonoids. *International Journal of Antimicrobial Agents* 26(5):343–356.

De Ledinghen V, Monvoisin A, Neaud V, Krisa S, Payrastre B, Bedin C, Desmouliere A, Bioulac-Sage P, Rosenbaum J (2001) Trans-resveratrol, a grapevine-derived polyphenol, blocks hepatocyte growth factor-induced invasion of hepatocellular carcinoma cells. *International Journal of Oncology* 19(1):83–88.

Ding Y, Zhao L, Mei H, Zhang S-L, Huang Z-H, Duan Y-Y, Ye P (2008) Exploration of Emodin to treat alpha-naphthylisothiocyanate-induced cholestatic hepatitis via anti-inflammatory pathway. *European Journal of Pharmacology* 590(1–3):377–386.

Dörrie J, Gerauer H, Wachter Y, Zunino SJ (2001) Resveratrol induces extensive apoptosis by depolarizing mitochondrial membranes and activating caspase-9 in acute lymphoblastic leukemia cells. *Cancer Research* 61(12):4731–4739.

Fanciulli M, Gentile FP, Bruno T, Paggi MG, Benassi M, Floridi A (1992) Inhibition of membrane redox activity by rhein and adriamycin in human glioma cells. *Anti-Cancer Drugs* 3(6):615–621.

Foust CM (2014) *Rhubarb: The Wondrous Drug*, Volume 191. Princeton, NJ: Princeton University Press.

Füllbeck M, Huang X, Dumdey R, Frommel C, Dubiel W, Preissner R (2005) Novel curcumin-and emodin-related compounds identified by in silico 2D/3D conformer screening induce apoptosis in tumor cells. *BMC Cancer* 5(1):97.

Gnanadesigan M, Ravikumar S, Anand M (2017) Hepatoprotective activity of Ceriops decandra (Griff.) Ding Hou mangrove plant against CCl 4 induced liver damage. *Journal of Taibah University for Science* 11(3):450–457.

Pharmacology

Goel V, Cheema SK, Agellon LB, Ooraikul B, Basu TK (1999) Dietary rhubarb (Rheum rhaponticum) stalk fibre stimulates cholesterol 7α-hydroxylase gene expression and bile acid excretion in cholesterol-fed C57BL/6J mice. *British Journal of Nutrition* 81(1):65–71.

Gupta RK, Bajracharya GB, Jha RN (2014) Antibacterial activity, cytotoxicity, antioxidant capacity and phytochemicals of Rheum australe rhizomes of Nepal. *Journal of Pharmacognosy and Phytochemistry* 2:125–128.

Hashemi M, Behrangi N, Borna H, Entezari M (2012) Protein tyrosine kinase (PTK) as a novel target for some natural anti-cancer molecules extracted from plants. *Journal of Medicinal Plants Research* 6(27):4375–4378.

Hatano T, Uebayashi H, Ito H, Shiota S, Tsuchiya T, Yoshida T (1999) Phenolic constituents of Cassia seeds and antibacterial effect of some naphthalenes and anthraquinones on methicillin-resistant Staphylococcus aureus. *Chemical and Pharmaceutical Bulletin* 47(8):1121–1127.

Hayashibara T, Yamada Y, Nakayama S, Harasawa H, Tsuruda K, Sugahara K, Miyanishi T, Kamihira S, Tomonaga M, Maita T (2002) Resveratrol induces downregulation in survivin expression and apoptosis in HTLV-1-infected cell lines: A prospective agent for adult T cell leukemia chemotherapy. *Nutrition and Cancer* 44(2):193–201.

He Z-H, He M-F, Ma S-C, But PP-H (2009) Anti-angiogenic effects of rhubarb and its anthraquinone derivatives. *Journal of Ethnopharmacology* 121(2):313–317.

Hilliard J, Krause H, Bernstein J, Fernandez J, Nguyen V, Ohemeng K, Barrett J (1995) A comparison of active site binding of 4-quinolones and novel flavone gyrase inhibitors to DNA gyrase. In: *Antimicrobial Resistance*. Berlin/Heidelberg, Germany: Springer:59–69.

Hosseini A, Mollazadeh H, Amiri MS, Sadeghnia HR, Ghorbani A (2017) Effects of a standardized extract of Rheum turkestanicum Janischew root on diabetic changes in the kidney, liver and heart of streptozotocin-induced diabetic rats. *Biomedicine and Pharmacotherapy* 86:605–611.

Hsu S-C, Chung J-G (2012) Anticancer potential of emodin. *Biology and Medicine* 2(3):108–116.

Hu B, Zhang H, Meng X, Wang F, Wang P (2014a) Aloe-emodin from rhubarb (Rheum rhabarbarum) inhibits lipopolysaccharide-induced inflammatory responses in RAW264. 7 macrophages. *Journal of Ethnopharmacology* 153(3):846–853.

Hu L (1986) Experimental study of Rheum officinale Baill. in treating severe hepatitis and hepatic coma. *Zhong xi yi jie he za zhi= Chinese Journal of Modern Developments in Traditional Medicine* 6(1):41.

Hu L, Chen N-N, Hu Q, Yang C, Yang Q-S, Wang F-F (2014b) An unusual piceatannol dimer from Rheum austral D. Don with antioxidant activity. *Molecules* 19(8):11453–11464.

Huang Q, Shen H-M, Ong C-N (2004) Inhibitory effect of emodin on tumor invasion through suppression of activator protein-1 and nuclear factor-κB. *Biochemical Pharmacology* 68(2):361–371.

Hussain H, Al-Harrasi A, Al-Rawahi A, Green IR, Csuk R, Ahmed I, Shah A, Abbas G, Rehman NU, Ullah R (2015) A fruitful decade from 2005 to 2014 for anthraquinone patents. *Expert Opinion on Therapeutic Patents* 25(9):1053–1064.

Ibrahim M, Anjum A, Waheed MA (2012) Curative effect of extracts of Sapindus mukorossi and Rheum emodi in CCl4 induced liver cirrhosis in male rats. *Global Journal of Medical Research* 12(8):47–52.

Ibrahim M, Khaja MN, Aara A, Khan AA, Habeeb MA, Devi YP, Narasu ML, Habibullah CM (2008) Hepatoprotective activity of Sapindus mukorossi and Rheum emodi extracts: In vitro and in vivo studies. *World Journal of Gastroenterology* 14(16):2566.

Ibrahim M, Khan AA, Tiwari SK, Habeeb MA, Khaja M, Habibullah C (2006) Antimicrobial activity of Sapindus mukorossi and Rheum emodi extracts against H pylori: In vitro and in vivo studies. *World Journal of Gastroenterology: WJG* 12(44):7136.

Iizuka A, Iijima OT, Kondo K, Itakura H, Yoshie F, Miyamoto H, Kubo M, Higuchi M, Takeda H, Matsumiya T (2004) Evaluation of Rhubarb using antioxidative activity as an index of pharmacological usefulness. *Journal of Ethnopharmacology* 91(1):89–94.

Ip S-W, Weng Y-S, Lin S-Y, Yang M-D, Tang N-Y, Su C-C, Chung J-G (2007) The role of Ca^{+2} on rhein-induced apoptosis in human cervical cancer Ca Ski cells. *Anticancer Research* 27(1A):379–389.

Jassim SAA, Naji MA (2003) Novel antiviral agents: A medicinal plant perspective. *Journal of Applied Microbiology* 95(3):412–427.

Jin B, Ma G, Wang H, Wang X (1989) Effects of rhubarb on electrical and contractive activities of the isolated intestine in rats. *Zhongguo zhong Yao za zhi= zhongguo zhongYao zazhi= China Journal of Chinese Materia Medica* 14(4):239–241, 256.

Jin H, Sakaida I, Tsuchiya M, Okita K (2005) Herbal medicine Rhei rhizome prevents liver fibrosis in rat liver cirrhosis induced by a choline-deficient L-amino acid-defined diet. *Life Sciences* 76(24):2805–2816.

Jong-Chol C, Tsukasa M, Kazuo A, Hiroaki K, Haruki Y, Yasuo O (1987) Anti-Bacteroides fragilis substance from rhubarb. *Journal of Ethnopharmacology* 19(3):279–283.

Kalcher K, Svancara I, Buzuk M, Vytras K, Walcarius A (2009) Electrochemical sensors and biosensors based on heterogeneous carbon materials. *Monatshefte für Chemie-Chemical Monthly* 140(8):861–889.

Kalisz S, Oszmiański J, Kolniak-Ostek J, Grobelna A, Kieliszek M, Cendrowski A (2020) Effect of a variety of polyphenols compounds and antioxidant properties of rhubarb (Rheum rhabarbarum). *LWT* 118:108775.

Kampa M, Hatzoglou A, Notas G, Damianaki A, Bakogeorgou E, Gemetzi C, Kouroumalis E, Martin P-M, Castanas E (2000) Wine antioxidant polyphenols inhibit the proliferation of human prostate cancer cell lines. *Nutrition and Cancer* 37(2):223–233.

Kang JH, Park YH, Choi SW, Yang EK, Lee WJ (2003) Resveratrol derivatives potently induce apoptosis in human promyelocytic leukemia cells. *Experimental and Molecular Medicine* 35(6):467–474.

Kang SC, Lee CM, Choung ES, Bak JP, Bae JJ, Yoo HS, Kwak JH, Zee OP (2008) Anti-proliferative effects of estrogen receptor-modulating compounds isolated from Rheum palmatum. *Archives of Pharmacal Research* 31(6):722–726.

Kaul TN, Middleton Jr E, Ogra PL (1985) Antiviral effect of flavonoids on human viruses. *Journal of Medical Virology* 15(1):71–79.

Khan SA, Ahmad A, Khan MI, Yusuf M, Shahid M, Manzoor N, Mohammad F (2012) Antimicrobial activity of wool yarn dyed with Rheum emodi L.(Indian Rhubarb). *Dyes and Pigments* 95(2):206–214.

Kim Y-W, Hyun C (2012) Evaluation of therapeutic effect of the extract from rhubarb (Rheum officinalis) in dogs with chronic renal failure. *Journal of Veterinary Clinics* 29(6):435–440.

Kimura Y, Baba K, Okuda H (2000) Inhibitory effects of active substances isolated from Cassia garrettiana heartwood on tumor growth and lung metastasis in Lewis lung carcinoma-bearing mice (Part 2). *Anticancer Research* 20(5A):2923–2930.

Kobori M, Masumoto S, Akimoto Y, Takahashi Y (2009) Dietary quercetin alleviates diabetic symptoms and reduces streptozotocin-induced disturbance of hepatic gene expression in mice. *Molecular Nutrition and Food Research* 53(7):859–868.

Pharmacology

Kounsar F, Afzal ZM (2010) Rheum emodi induces nitric oxide synthase activity in murine macrophages. *American Journal of Biomedical Sciences* 2(2):155–163.

Kounsar F, Rather MA, Ganai BA, Zargar MA (2011) Immuno-enhancing effects of the herbal extract from Himalayan rhubarb Rheum emodi Wall. ex Meissn. *Food Chemistry* 126(3):967–971.

Kumar DN, George VC, Suresh P, Kumar RA (2013) Acceleration of pro-caspase-3 maturation and cell migration inhibition in human breast cancer cells by phytoconstituents of Rheum emodi rhizome extracts. *Excli Journal* 12:462.

Kumar DN, Shikha DS, George VC, Suresh P, Kumar RA (2012) Anticancer and antimetastatic activities of Rheum emodi rhizome chloroform extracts. *Asian Journal of Pharmaceutical and Clinical Research* 5(3):189–194.

Kumar N, Ragupathi D, George VC, Suresh PK, Kumar RA (2015) Cancer-specific chemoprevention and anti-metastatic potentials of Rheum emodi rhizome ethyl acetate extracts and identification of active principles through HPLC and GC-MS analysis. *Pakistan Journal of Pharmaceutical Sciences* 28(1):83–93.

Kuo P-L, Lin T-C, Lin C-C (2002) The antiproliferative activity of aloe-emodin is through p53-dependent and p21-dependent apoptotic pathway in human hepatoma cell lines. *Life Sciences* 71(16):1879–1892.

Kurokawa M, Ochiai H, Nagasaka K, Neki M, Xu H, Kadota S, Sutardjo S, Matsumoto T, Namba T, Shiraki K (1993) Antiviral traditional medicines against herpes simplex virus (HSV-1), poliovirus, and measles virus in vitro and their therapeutic efficacies for HSV-1 infection in mice. *Antiviral Research* 22(2–3):175–188.

Larrosa M, Tomás-Barberán FA, Espín JC (2004) The grape and wine polyphenol piceatannol is a potent inducer of apoptosis in human SK-Mel-28 melanoma cells. *European Journal of Nutrition* 43(5):275–284.

Lee H-Z, Hsu S-L, Liu M-C, Wu C-H (2001) Effects and mechanisms of aloe-emodin on cell death in human lung squamous cell carcinoma. *European Journal of Pharmacology* 431(3):287–295.

Lee MS, Sohn CB (2008) Anti-diabetic properties of chrysophanol and its glucoside from rhubarb rhizome. *Biological and Pharmaceutical Bulletin* 31(11):2154–2157.

Leng-Peschlow E (1986) Dual effect of orally administered Sennosides on large intestine transit and fluid absorption in the rat. *Journal of Pharmacy and Pharmacology* 38(8):606–610.

Lewis W, Elvin-Lewis M (1977) Internal poisons. In: *Medical Botany: Plants Affecting Man's Health*. New York: John Wiley:11–63.

Li C-Y, Li X, Liu L-J, Zou J, Hu F, Li J-Y (2008) Effect of aloe-emodin on proliferation and [Ca~(2+)] _i mobilization of T lymphocytes. *Chinese Traditional and Herbal Drugs* 8:1192–1196.

Li J, Li G, Miao Y, Wu Z, Li L (2009) Therapeutic effect of Aloe-emodin for experimental acute pancreatitis in rats. *China med Her* 6:14–15.

Lin C-W, Wu C-F, Hsiao N-W, Chang C-Y, Li S-W, Wan L, Lin Y-J, Lin W-Y (2008) Aloe-emodin is an interferon-inducing agent with antiviral activity against Japanese encephalitis virus and enterovirus 71. *International Journal of Antimicrobial Agents* 32(4):355–359.

Lin H-Y, Shih A, Davis FB, Tang H-Y, Martino LJ, Bennett JA, Davis PJ (2002) Resveratrol induced serine phosphorylation of p53 causes apoptosis in a mutant p53 prostate cancer cell line. *The Journal of Urology* 168(2):748–755.

Lin J-G, Chen G-W, Li T-M, Chouh S-T, Tan T-W, Chung J-G (2006) Aloe-emodin induces apoptosis in T24 human bladder cancer cells through the p53 dependent apoptotic pathway. *The Journal of Urology* 175(1):343–347.

Lin M-L, Lu Y-C, Chung J-G, Li Y-C, Wang S-G, Sue-Hwee N, Wu C-Y, Su H-L, Chen S-S (2010) Aloe-emodin induces apoptosis of human nasopharyngeal carcinoma cells via caspase-8-mediated activation of the mitochondrial death pathway. *Cancer Letters* 291(1):46–58.

Litescu SC, Eremia SA, Diaconu M, Tache A, Radu G-L (2011) Biosensors applications on assessment of reactive oxygen species and antioxidants. *Environmental Biosensors*18:95–114.

Liu S, Sporer F, Wink M, Jourdane J, Henning R, Li Y, Ruppel A (1997) Anthraquinones in Rheum palmatum and Rumex dentatus (Polygonaceae), and phorbol esters in Jatropha curcas (Euphorbiaceae) with molluscicidal activity against the schistosome vector snails Oncomelania, Biomphalaria, and Bulinus. *Tropical Medicine and International Health : TM and IH* 2(2):179–188.

Liu Y, Yan F, Liu Y, Zhang C, Yu H, Zhang Y, Zhao Y (2008) Aqueous extract of rhubarb stabilizes vulnerable atherosclerotic plaques due to depression of inflammation and lipid accumulation. *Phytotherapy Research: PTR* 22(7):935–942.

Liu Z, Ma N, Zhong Y, Yang Z-q (2015) Antiviral effect of emodin from Rheum palmatum against coxsakievirus B 5 and human respiratory syncytial virus in vitro. *Journal of Huazhong University of Science and Technology [Medical Sciences]* 35(6):916–922.

López-Lázaro M, Willmore E, Austin CA (2010) The dietary flavonoids myricetin and fisetin act as dual inhibitors of DNA topoisomerases I and II in cells. *Mutation Research/Genetic Toxicology and Environmental Mutagenesis* 696(1):41–47.

Lopez-Novoa JM, Quiros Y, Vicente L, Morales AI, Lopez-Hernandez FJ (2011) New insights into the mechanism of aminoglycoside nephrotoxicity: An integrative point of view. *Kidney International* 79(1):33–45.

Lu C, Wang H, Lv W, Xu P, Zhu J, Xie J, Liu B, Lou Z (2011) Antibacterial properties of anthraquinones extracted from rhubarb against Aeromonas hydrophila. *Fisheries Science* 77(3):375.

Lu CC, Yang JS, Huang AC, Hsia TC, Chou ST, Kuo CL, Lu HF, Lee TH, Wood WG, Chung JG (2010) Chrysophanol induces necrosis through the production of ROS and alteration of ATP levels in J5 human liver cancer cells. *Molecular Nutrition and Food Research* 54(7):967–976.

Lu J, Papp LV, Fang J, Rodriguez Nieto S, Zhivotovsky B, Holmgren A (2006) Inhibition of mammalian thioredoxin reductase by some flavonoids: Implications for myricetin and quercetin anticancer activity. *Cancer Research* 66(8):4410–4418.

Mahyar-Roemer M, Köhler H, Roemer K (2002) Role of Bax in resveratrol-induced apoptosis of colorectal carcinoma cells. *BMC Cancer* 2(1):27.

Martinez-Salgado C, López-Hernández FJ, López-Novoa JM (2007) Glomerular nephrotoxicity of aminoglycosides. *Toxicology and Applied Pharmacology* 223(1):86–98.

May G, Willuhn G (1978) Anti-viral activity of aqueous extracts from medicinal-plants in tissue-cultures. *Arzneimittel-Forschung/Drug Research* 28(1):1–7.

Meng Z, Yan Y, Tang Z, Guo C, Li N, Huang W, Ding G, Wang Z, Xiao W, Yang Z (2015) Anti-hyperuricemic and nephroprotective effects of rhein in hyperuricemic mice. *Planta Medica* 81(4):279–285

Mgbonyebi OP, Russo J, Russo IH (1998) Antiproliferative effect of synthetic resveratrol on human breast epithelial cells. *International Journal of Oncology* 12(4):865–874.

Miraj S (2016) Therapeutic effects of Rheum palmatum L.(Dahuang): A systematic review. *Der Pharma Chemica* 8(13):50–54.

Mirzoeva O, Grishanin R, Calder P (1997) Antimicrobial action of propolis and some of its components: The effects on growth, membrane potential and motility of bacteria. *Microbiological Research* 152(3):239–246.

Mishra SK, Tiwari S, Shrivastava A, Srivastava S, Boudh GK, Chourasia SK, Chaturvedi U, Mir SS, Saxena AK, Bhatia G, Lakshmi V (2014) Antidyslipidemic effect and antioxidant activity of anthraquinone derivatives from Rheum emodi rhizomes in dyslipidemic rats. *Journal of Natural Medicines* 68(2):363–371.

Misiti F, Sampaolese B, Mezzogori D, Orsini F, Pezzotti M, Giardina B, Clementi M (2006) Protective effect of rhubarb derivatives on amyloid beta (1–42) peptide-induced apoptosis in IMR-32 cells: A case of nutrigenomic. *Brain Research Bulletin* 71(1–3):29–36.

Molyneux P (2004) The use of the stable free radical diphenylpicrylhydrazyl (DPPH) for estimating antioxidant activity. *Songklanakarin Journal of Science and Technology* 26(2):211–219.

Moon MK, Kang DG, Lee JK, Kim JS, Lee HS (2006) Vasodilatory and anti-inflammatory effects of the aqueous extract of rhubarb via a NO-cGMP pathway. *Life Sciences* 78(14):1550–1557.

Mori A, Nishino C, Enoki N, Tawata S (1987) Antibacterial activity and mode of action of plant flavonoids against Proteus vulgaris and Staphylococcus aureus. *Phytochemistry* 26(8):2231–2234.

Naqishbandi AM, Adham AN (2015) HPLC analysis and antidiabetic effect of Rheum ribes root in type 2 diabetic patients. *Zanco Journal of Medical Sciences* 19(2):957–964.

Neyrinck AM, Etxeberria U, Jouret A, Delzenne NM (2014) Supplementation with crude rhubarb extract lessens liver inflammation and hepatic lipid accumulation in a model of acute alcohol-induced steato-hepatitis. *Archives of Public Health* 72(S1):6.

Neyrinck AM, Etxeberria U, Taminiau B, Daube G, Van Hul M, Everard A, Cani PD, Bindels LB, Delzenne NM (2017) Rhubarb extract prevents hepatic inflammation induced by acute alcohol intake, an effect related to the modulation of the gut microbiota. *Molecular Nutrition and Food Research* 61(1):1500899.

Ngai HH, Sit W-H, Wan JM (2005) The nephroprotective effects of the herbal medicine preparation, WH30+, on the chemical-induced acute and chronic renal failure in rats. *The American Journal of Chinese Medicine* 33(3):491–500.

Niles RM, McFarland M, Weimer MB, Redkar A, Fu Y-M, Meadows GG (2003) Resveratrol is a potent inducer of apoptosis in human melanoma cells. *Cancer Letters* 190(2):157–163.

Öztürk M, Aydoğmuş-Öztürk F, Duru ME, Topçu G (2007) Antioxidant activity of stem and root extracts of Rhubarb (Rheum ribes): An edible medicinal plant. *Food Chemistry* 103(2):623–630.

Pandith SA, Dar RA, Lattoo SK, Shah MA, Reshi ZA (2018) Rheum australe, an endangered high-value medicinal herb of North Western Himalayas: A review of its botany, ethnomedical uses, phytochemistry and pharmacology. *Phytochemistry Reviews : Proceedings of the Phytochemical Society of Europe* 17(3):573–609.

Pandith SA, Hussain A, Bhat WW, Dhar N, Qazi AK, Rana S, Razdan S, Wani TA, Shah MA, Bedi Y, Hamid A, Lattoo SK (2014) Evaluation of anthraquinones from Himalayan rhubarb (Rheum emodi Wall. ex Meissn.) as antiproliferative agents. *South African Journal of Botany* 95:1–8.

Papke A, Möller F, Rettenberger R, Kaszkin-Bettag M, Vollmer G (2008) The special extract of Rheum rhaponticum (ERr 731®) does not stimulate uterine growth and proliferation in ovariectomized Wistar rats. *Zeitschrift für Phytotherapie* 29(S1):P28.

Pecere T, Gazzola MV, Mucignat C, Parolin C, Dalla Vecchia F, Cavaggioni A, Basso G, Diaspro A, Salvato B, Carli M, Palù G (2000) Aloe-emodin is a new type of anticancer agent with selective activity against neuroectodermal tumors. *Cancer Research* 60(11):2800–2804.

Pisoschi AM, Negulescu GP (2011) Methods for total antioxidant activity determination: A review. *Biochemistry and Analytical Biochemistry* 1(1):106.

Plaper A, Golob M, Hafner I, Oblak M, Šolmajer T, Jerala R (2003) Characterization of quercetin binding site on DNA gyrase. *Biochemical and Biophysical Research Communications* 306(2):530–536.

Raal A, Pokk P, Arend A, Aunapuu M, Jõgi J, Õkva K, Püssa T (2009) *trans*-resveratrol alone and hydroxystilbenes of rhubarb (*Rheum rhaponticum* L.) root reduce liver damage induced by chronic ethanol administration: A comparative study in mice. *Phytotherapy Research: An International Journal Devoted to Pharmacological and Toxicological Evaluation of Natural Product Derivatives* 23(4):525–532.

Rafiq RA, Quadri A, Nazir LA, Peerzada K, Ganai BA, Tasduq SA (2015) A potent inhibitor of phosphoinositide 3-kinase (PI3K) and mitogen activated protein (MAP) kinase signalling, quercetin (3, 3', 4', 5, 7-pentahydroxyflavone) promotes cell death in ultraviolet (UV)-B-irradiated B16F10 melanoma cells. *PLOS ONE* 10(7):e0131253.

Rajkumar V, Guha G, Ashok Kumar R (2011a) Antioxidant and anti-cancer potentials of Rheum emodi rhizome extracts. *Evidence-Based Complementary and Alternative Medicine* (https://doi.org/10.1093/ecam/neq048)..

Rajkumar V, Guha G, Kumar RA (2011b) Apoptosis induction in MDA-MB-435S, Hep3B and PC-3 cell lines by Rheum emodi rhizome extracts. *Asian Pacific Journal of Cancer Prevention: APJCP* 12(5):1197–1200.

Ravishankar D, Rajora AK, Greco F, Osborn HM (2013) Flavonoids as prospective compounds for anti-cancer therapy. *The International Journal of Biochemistry and Cell Biology* 45(12):2821–2831.

Ristow M (2014) Unraveling the truth about antioxidants: Mitohormesis explains ROS-induced health benefits. *Nature Medicine* 20(7):709–711.

Rokaya MB, Münzbergová Z, Timsina B, Bhattarai KR (2012) Rheum australe D. Don: A review of its botany, ethnobotany, phytochemistry and pharmacology. *Journal of Ethnopharmacology* 141(3):761–774.

Roupe KA, Helms GL, Halls SC, Yáñez JA, Davies NM (2005) Preparative enzymatic synthesis and HPLC analysis of rhapontigenin: Applications to metabolism, pharmacokinetics and anti-cancer studies. *Journal of Pharmacy and Pharmaceutical Sciences : A Publication of the Canadian Society for Pharmaceutical Sciences, Societe Canadienne des Sciences Pharmaceutiques* 8(3):374–386.

Roupe KA, Remsberg CM, Yáñez JA, Davies NM (2006) Pharmacometrics of stilbenes: Seguing towards the clinic. *Current Clinical Pharmacology* 1(1):81–101.

Saito ST, Silva G, Pungartnik C, Brendel M (2012) Study of DNA–emodin interaction by FTIR and UV–vis spectroscopy. *Journal of Photochemistry and Photobiology, Part B: Biology* 111:59–63.

Sales MDC, Costa HB, Fernandes PMB, Ventura JA, Meira DD (2016) Antifungal activity of plant extracts with potential to control plant pathogens in pineapple. *Asian Pacific Journal of Tropical Biomedicine* 6(1):26–31.

Scarlatti F, Maffei R, Beau I, Codogno P, Ghidoni R (2008) Role of non-canonical Beclin 1-independent autophagy in cell death induced by resveratrol in human breast cancer cells. *Cell Death and Differentiation* 15(8):1318–1329.

Semple SJ, Pyke SM, Reynolds GD, Flower RL (2001) In vitro antiviral activity of the anthraquinone chrysophanic acid against poliovirus. *Antiviral Research* 49(3):169–178.

Serrero G, Lu R (2001) Effect of resveratrol on the expression of autocrine growth modulators in human breast cancer cells. *Antioxidants and Redox Signaling* 3(6):969–979.

Pharmacology 103

Shen M-Y, Lin Y-P, Yang B-C, Jang Y-S, Chiang C-K, Mettling C, Chen Z-W, Sheu J-R, Chang CL, Lin Y-L, Yang WC (2012) Catenarin prevents type 1 diabetes in non-obese diabetic mice via inhibition of leukocyte migration involving the MEK6/p38 and MEK7/JNK pathways. *Evidence-Based Complementary and Alternative Medicine* (https://doi.org/10.1155/2012/982396).

Shia C-S, Juang S-H, Tsai S-Y, Chang P-H, Kuo S-C, Hou Y-C, Chao P-DL (2009) Metabolism and pharmacokinetics of anthraquinones in Rheum palmatum in rats and ex vivo antioxidant activity. *Planta Medica* 75(13):1386–1392.

Shukla SD, Pruett SB, Szabo G, Arteel GE (2013) Binge ethanol and liver: New molecular developments. *Alcoholism: Clinical and Experimental Research* 37(4):550–557.

Srinivas G, Anto RJ, Srinivas P, Vidhyalakshmi S, Senan VP, Karunagaran D (2003) Emodin induces apoptosis of human cervical cancer cells through poly (ADP-ribose) polymerase cleavage and activation of caspase-9. *European Journal of Pharmacology* 473(2–3):117–125.

Srinivas G, Babykutty S, Sathiadevan PP, Srinivas P (2007) Molecular mechanism of emodin action: Transition from laxative ingredient to an antitumor agent. *Medicinal Research Reviews* 27(5):591–608.

Srinivasarao M, Mangamoori Lakshminarasu TM, Ibrahim M (2015) *Hepatoprotective Potential of Total Anthraquinone Fraction of Rheum Emodi: In Vitro and In Vivo Studies. Journal of Medical and Pharmaceutical Innovation* 2(11):7–14.

Stewart JR, O'Brian CA (2004) Resveratrol antagonizes EGFR-dependent ERK1/2 activation in human androgen-independent prostate cancer cells with associated isozyme-selective PKCα inhibition. *Investigational New Drugs* 22(2):107–117.

Suboj P, Babykutty S, Gopi DRV, Nair RS, Srinivas P, Gopala S (2012) Aloe emodin inhibits colon cancer cell migration/angiogenesis by downregulating MMP-2/9, RhoB and VEGF via reduced DNA binding activity of NF-κB. *European Journal of Pharmaceutical Sciences: Official Journal of the European Federation for Pharmaceutical Sciences* 45(5):581–591.

Sydiskis R, Owen D, Lohr J, Rosler K, Blomster R (1991) Inactivation of enveloped viruses by anthraquinones extracted from plants. *Antimicrobial Agents and Chemotherapy* 35(12):2463–2466.

Tahir M, Qamar MA, ul Haq I, Malik MM, Rauf S (2009) Inhibitory effect of 'Rheum emodi Wall' on hepatic cytochrome P450 enzymes. *Pakistan Journal of Physiology* 5(1):73–75.

Thangaraj P (2016) Anti-inflammatory activity. In: *Pharmacological Assays of Plant-Based Natural Products*, Volume 7. Berlin/Heidelberg, Germany: Springer:103–111.

Tsan MF, White JE, Maheshwari JG, Bremner TA, Sacco J (2000) Resveratrol induces Fas signalling-independent apoptosis in THP-1 human monocytic leukaemia cells. *British Journal of Haematology* 109(2):405–412.

Usui T, Ikeda Y, Tagami T, Matsuda K, Moriyama K, Yamada K, Kuzuya H, Kohno S, Shimatsu A (2002) The phytochemical lindleyin, isolated from Rhei rhizoma, mediates hormonal effects through estrogen receptors. *Journal of Endocrinology* 175(2):289–296.

Uyar P, Coruh N, İscan M (2014) Evaluation of in vitro antioxidative, cytotoxic and apoptotic activities of Rheum ribes ethyl acetate extracts. *Journal of Plant Sciences* 2(6):339–346.

Waffo-Téguo P, Hawthorne ME, Cuendet M, Mérillon J-M, Kinghorn AD, Pezzuto JM, Mehta RG (2001) Potential cancer-chemopreventive activities of wine stilbenoids and flavans extracted from grape (Vitis vinifera) cell cultures. *Nutrition and Cancer* 40(2):173–179.

Waly MI, Ali BH, Al-Lawati I, Nemmar A (2013) Protective effects of emodin against cisplatin-induced oxidative stress in cultured human kidney (HEK 293) cells. *Journal of Applied Toxicology : JAT* 33(7):626–630.

Wang H, Feng F, Zhuang B-Y, Sun Y (2009) Evaluation of hepatoprotective effect of Zhi-Zi-Da-Huang decoction and its two fractions against acute alcohol-induced liver injury in rats. *Journal of Ethnopharmacology* 126(2):273–279.

Wang J-B, Zhao H-P, Zhao Y-L, Jin C, Liu D-J, Kong W-J, Fang F, Zhang L, Wang H-J, Xiao X-H (2011) Hepatotoxicity or hepatoprotection? Pattern recognition for the paradoxical effect of the Chinese herb Rheum palmatum L. in treating rat liver injury. *PLOS ONE* 6(9):e24498.

Wang J, Liu S, Yin Y, Li M, Wang B, Yang L, Jiang Y (2015) FOXO3-mediated upregulation of Bim contributes to rhein-induced cancer cell apoptosis. *Apoptosis : An International Journal on Programmed Cell Death* 20(3):399–409.

Wang Z, Song H (1999) Clinical observation on therapeutical effect of prepared rhubarb in treating pregnancy induced hypertension. *Zhongguo Zhong Xi Yi Jie He Za Zhi Zhongguo Zhongxiyi Jiehe Zazhi= Chinese Journal of Integrated Traditional and Western Medicine* 19(12):725–727.

Wang Z, Wang G, Xu H, Wang P (1996) Anti-herpes virus action of ethanol-extract from the root and rhizome of Rheum officinale Baill. *Zhongguo zhong Yao za zhi= zhongguo zhongYao zazhi= China Journal of Chinese Materia Medica* 21(6):364–366, 384.

Wolter F, Clausnitzer A, Akoglu B, Stein J (2002) Piceatannol, a natural analog of resveratrol, inhibits progression through the S phase of the cell cycle in colorectal cancer cell lines. *The Journal of Nutrition* 132(2):298–302.

Xing X-Y, Zhao Y-L, Kong W-J, Wang J-B, Jia L, Zhang P, Yan D, Zhong Y-W, Li R-S, Xiao X-H (2011) Investigation of the "dose–time–response" relationships of rhubarb on carbon tetrachloride-induced liver injury in rats. *Journal of Ethnopharmacology* 135(2):575–581.

Xiong H-R, Luo J, Hou W, Xiao H, Yang Z-Q (2011) The effect of emodin, an anthraquinone derivative extracted from the roots of Rheum tanguticum, against herpes simplex virus in vitro and in vivo. *Journal of Ethnopharmacology* 133(2):718–723.

Xu X, Lv H, Xia Z, Fan R, Zhang C, Wang Y, Wang D (2017) Rhein exhibits antioxidative effects similar to Rhubarb in a rat model of traumatic brain injury. *BMC Complementary and Alternative Medicine* 17(1):1–9.

Yamauchi K, Yagi T, Kuwano S (1993) Suppression of the purgative action of rhein anthrone, the active metabolite of Sennosides A and B, by calcium channel blockers, calmodulin antagonists and indometacin. *Pharmacology* 47 (S1):22–31.

Yang J, Li L (1993) Effects of Rheum on renal hypertrophy and hyperfiltration of experimental diabetes in rat. *Zhongguo Zhong Xi Yi Jie He Za Zhi Zhongguo Zhongxiyi Jiehe Zazhi= Chinese Journal of Integrated Traditional and Western Medicine* 13(5):286–288, 261–282.

Yang M, Li X, Zeng X, Ou Z, Xue M, Gao D, Liu S, Li X, Yang S (2016) Rheum palmatum L. attenuates high fat diet-induced hepatosteatosis by activating AMP-activated protein kinase. *The American Journal of Chinese Medicine* 44(3):551–564.

Yang SH, Kim JS, Oh TJ, Kim MS, Lee SW, Woo SK, Cho HS, Choi YH, Kim YH, Rha SY, Chung HC, An SW (2003a) Genome-scale analysis of resveratrol-induced gene expression profile in human ovarian cancer cells using a cDNA microarray. *International Journal of Oncology* 22(4):741–750.

Yang W, LI X, Liu B, Chen L, Wang H, Zhao Y, QI Q (2003b) The inhibition of rhein on increase. In: [ca~(2+)] i and TNFα release in lipopolysaccharide-stimulated macrophages. *Acta Scientiarun Naturaltium Universitatis Nankaiensis* 3:111–115.

Yen G-C, Duh P-D, Chuang D-Y (2000) Antioxidant activity of anthraquinones and anthrone. *Food Chemistry* 70(4):437–441.

Yim H, Lee YH, Lee CH, Lee SK (1999) Emodin, an anthraquinone derivative isolated from the rhizomes of Rheum palmatum, selectively inhibits the activity of casein kinase II as a competitive inhibitor. *Planta Medica* 65(1):9–13.

Yin J, Zhang A, Wang Y, Zhang H (1993) Anti-inflammatory mechanism of rhein-arginine in preventing rats ankylenteron. *Traditional Chinese Drug Research and Clinical Pharmacology* 1:23–28.

Youl E, Bardy G, Magous R, Cros G, Sejalon F, Virsolvy A, Richard S, Quignard J, Gross R, Petit P, Bataille D, Oiry C (2010) Quercetin potentiates insulin secretion and protects INS-1 pancreatic β-cells against oxidative damage via the ERK1/2 pathway. *British Journal of Pharmacology* 161(4):799–814.

Yu M, Luo Y-L, Zheng J-W, Ding Y-H, Li W, Zheng T-Z, Qu S-Y (2005) Effects of rhubarb on isolated gastric muscle strips of guinea pigs. *World Journal of Gastroenterology: WJG* 11(17):2670.

Zargar BA, Masoodi MH, Ahmed B, Ganie SA (2011) Phytoconstituents and therapeutic uses of Rheum emodi wall. ex Meissn. *Food Chemistry* 128(3):585–589.

Zhang L, Lau Y-K, Xi L, Hong R-L, Kim DS, Chen C-F, Hortobagyi GN, Chang C-j, Hung M-C (1998) Tyrosine kinase inhibitors, emodin and its derivative repress HER-2/neu-induced cellular transformation and metastasis-associated properties. *Oncogene* 16(22):2855–2863.

Zhang R, Kang KA, Piao MJ, Lee KH, Jang HS, Park MJ, Kim BJ, Kim JS, Kim YS, Ryu SY, Hyun JW (2007) Rhapontigenin from Rheum undulatum protects against oxidative-stress-induced cell damage through antioxidant activity. *Journal of Toxicology and Environmental Health. Part A* 70(13):1155–1166.

Zhang T, Wu Z, Du J, Hu Y, Liu L, Yang F, Jin Q (2012) Anti-Japanese-encephalitis-viral effects of kaempferol and daidzin and their RNA-binding characteristics. *PLOS ONE* 7(1):e30259.

Zhang Y, Liu D (2011) Flavonol kaempferol improves chronic hyperglycemia-impaired pancreatic beta-cell viability and insulin secretory function. *European Journal of Pharmacology* 670(1):325–332.

Zhang Y, Ma H, Mai X, Xu Z, Yang Y, Wang H, Ouyang L, Liu S (2019a) Comparative pharmacokinetics and metabolic profile of Rhein following oral administration of Niuhuang Shang Qing tablets, rhubarb and Rhein in rats. *International Journal of Pharmacology* 15(1):19–30.

Zhang Y, Ouyang L, Mai X, Wang H, Liu S, Zeng H, Chen T, Li J (2019b) Use of UHPLC-QTOF-MS/MS with combination of in silico approach for distributions and metabolites profile of flavonoids after oral administration of Niuhuang Shangqing tablets in rats. *Journal of Chromatography. Part B, Biomedical Sciences and Applications* 1114:55–70.

Zhang Z-H, Vaziri ND, Wei F, Cheng X-L, Bai X, Zhao Y-Y (2016) An integrated lipidomics and metabolomics reveal nephroprotective effect and biochemical mechanism of Rheum officinale in chronic renal failure. *Scientific Reports* 6:22151.

Zhang Z-H, Wei F, Vaziri ND, Cheng X-L, Bai X, Lin R-C, Zhao Y-Y (2015) Metabolomics insights into chronic kidney disease and modulatory effect of rhubarb against tubulointerstitial fibrosis. *Scientific Reports* 5:14472.

Zhang Z, Cheng W, Yang Y (1994) Low-dose of processed rhubarb in preventing pregnancy induced hypertension. *Zhonghua Fu Chan ke za zhi* 29(8):463–464, 509.

Zheng JM, Zhu JM, Li LS, Liu ZH (2008) Rhein reverses the diabetic phenotype of mesangial cells over-expressing the glucose transporter (GLUT1) by inhibiting the hexosamine pathway. *British Journal of Pharmacology* 153(7):1456–1464.

Zheng Q-x, Wu H-f, Jian G, Nan H-j, Chen S-l, Yang J-s, Xu X-d (2013) Review of rhubarbs: Chemistry and pharmacology. *Chinese Herbal Medicines* 5(1):9–32.

Zhuang S, Yu R, Zhong J, Liu P, Liu Z (2019) Rhein from Rheum rhabarbarum inhibits hydrogen-peroxide-induced oxidative stress in intestinal epithelial cells partly through PI3K/Akt-mediated Nrf2/HO-1 pathways. *Journal of Agricultural and Food Chemistry* 67(9):2519–2529.

6 Molecular Aspects

6.1 CYTOGENETICS

The nucleotide composition, physicochemical structure (numerous elements organized in the genome in many ways), and functions of chromatin (the non-coding constitutive heterochromatin and the coding euchromatin containing unique sequences and active genes) are highly variable. The chromatin holds importance in systematics as it preserves its specific physicochemical pattern (expressed in unwavering distribution of its types in the cell nucleus and on the mitotic chromosomes in the form of particular segments called bands) from one generation to the next. In fact, a species differs from every other species in the quantity, composition, and distribution of its nucleotides, and it is on this basis we can place the organisms in different taxonomic ranks within a family, genus, or even a species, viz. subspecies, variety, or cytotype (Ilnicki 2014). The classification of organisms based on cytological characters, i.e., karyosystematics (Joachimiak et al. 1997) or cytotaxonomy (Peruzzi and Eroğlu 2013) includes relative studies of karyological features which are then used in plant cytogenetics (the study of inheritance in relation to the structure and function of chromosomes) to elucidate the evolutionary patterns within species/genera/families vis-à-vis migration and colonization of species to different ecological niches (Siljak-Yakovlev and Peruzzi 2012; Te Beest et al. 2012). Indeed, the cytogenetic data is important in investigating the processes of diversification and evolution in plants, particularly for the taxa distributed in the arid habitats. Pertinently, decades of classical research in karyology (and karyotyping) has resulted in reasonable understanding about evolutionary processes taking place in both plants and animals. Based on karyological data, the remarkable accomplishments in plant biosystematics include phylogenetic division of the families Poaceae and Ranunculaceae, and formation of the family Agavaceae from the genera *Agave* and *Yucca* containing a bimodal karyotype. Further, notwithstanding the visible differences in the floral structure of Cyperaceae and Juncaceae, it advocates a close relationship between these two families on the basis of the existence of infrequent holokinetic chromosomes and reverse meiosis (Gregory 1941; Stace 1991). Moreover, events of hybridization (process of interbreeding between genetically divergent individuals, responsible for reticulate evolution) and polyploidization (possession of three or more basic chromosome sets in nuclei) exhibit common occurrence in plants in natural habitats (Ilnicki and Szelag 2011; Te Beest et al. 2012). Both these phenomena are prevalent in many angiosperm species (Wissemann 2007) and have been known to be the prominent driving force for plant diversification and speciation episodes assisting species to occupy new territories (Ruirui et al. 2010; Te Beest et al. 2012). Studies suggest that traditional ways of working with chromosomes should

DOI: 10.1201/9780429340390-6

be used as supplementation to the contemporary methods of cytogenetics (FISH, GISH, BAC, chromosome painting, etc.) to recognize the molecular structure of chromatin and chromosomes (Guerra 2012).

It has been reported that the genus *Rheum* is monobasic with all species having x = 11 (Fedorov 1969; Darlington and Wylie 1956). Nearly 40% of the cytologically known species of *Rheum* are tetraploid, 48% diploid, and the remaining 12% have both diploid and tetraploid cytotypes (Saggoo and Farooq 2011; Ruirui et al. 2010). However, and to date, only diploid cytotypes have been reported in one of the most important species of this genus, *R. australe* bearing chromosome numbers 2n = 22 (Gohil and Rather 1986; Saggoo and Farooq 2011).

In an earlier investigation Chin and Youngken attempted to clarify the taxonomic issues in genus *Rheum* wherein a preliminary survey on taxonomy, chemistry, ecology, and cytology of various species of *Rheum* was conducted. The diploid numbers of chromosomes were found to be 22 for *R. franzenbachii*, *R. undulatum*, and *R. emodi*, and 44 for *R. tataricum*, *R. compactum*, *R. altaicum*, *R. webbianum*, *R. rhaponticum*, and *R. australe*. The diploid species (2n = 22) were found to have chromosomes of similar size, though differing in their morphology which made the authors to divide these species into two apparent groups, viz. "*palmatum-franzenbachii*" type (with the longest pair of chromosomes having sub-median centromeres, and a pair of satellite chromosomes) and "*emodi-undulatum*" type (longest pair possess sub-terminal centromeres, and two pairs of chromosomes with satellites) (Chin and Youngken 1947). Although *R. emodi* is still being used as a synonym for *R. australe*, recent literature considers the latter to be the correct (legitimate) name of the species (Pandith et al. 2018). The rhubarb cultivars are mostly tetraploid (2n = 44) (Libert 1987).

Polyploidy, the driving force for diversification and speciation (particularly in alpine mountains) experienced by nearly 80% of angiosperm species, is relatively known to exhibit less frequency in some of the species-rich genera in the QTP (China) and its adjoining regions (Ruirui et al. 2010; Liu 2004; Nie et al. 2005). For instance, one such genus *Rheum* is represented by a mere two ploidy levels, viz. diploid and tetraploid (Darlington and Janaki Ammal 1945; Janaki Ammal 1955). Chromosome size, number, and karyotype are well known taxonomic attributes for cytological studies focused on diversification. Indeed, karyotype is an important parameter that allows the authentic identification of a species (Stebbins 1971). Taking this as background, Ruirui et al. (2010) selected 12 morphologically diversified species of *Rheum* representing five sections and aimed (1) to contribute more chromosomal data to the genus *Rheum* and to know whether polyploids occur in more species, and (2) to examine possible karyotypic differentiations within selected divergent congeneric species. Besides confirming the earlier reported basic chromosome number (x = 11) and ploidy levels (diploid and tetraploid), the authors found nine (*R. webbianum*, *R. likiangense*, *R. palmatum*, *R. tanguticum*, *R. nanum*, *R. rhomboideum*, *R. reticulatum*, *R. nobile*, and *R. alexandrae*) species as diploid and the rest three (*R. wittrockii*, *R. compactum*, and *R. pumilum*) as tetraploid. Certainly, the authors report chromosome numbers and karyotypes for the first time for seven species among the chosen ones

Molecular Aspects

and include *R. alexandrae, R. likiangense, R. nanum, R. nobile, R. pumilum, R. reticulatum*, and *R. rhomboideum*. Karyotyping showed that the chromosomes were small in size with no satellites. Besides, the karyotypes contained metacentric as well as a few sub-metacentric chromosomes in both diploid and tetraploid species. Pertinently, the study concluded that karyotypic differentiation has a lower share in inter-specific morphological divergence within the genus *Rheum*. Nonetheless, the role of polyploidy in its diversification (and further speciation) cannot be neglected.

A symmetrical karyotype (chromosome morphology) has chromosomes of similar size with the centromere in the middle (or slightly dislocated) position whereas an asymmetrical (or heterogeneous) karyotype results from the shift of the centromere position to the terminal side or through the accumulation of differences in the relative size between the chromosome complement. The symmetric karyotypes are mostly found in primitive or primordial species, whereas the advanced plant taxa show asymmetric karyotypes. Generally, in higher plants, karyotypes are known to switch from symmetry to asymmetry. They exhibit variability in five unlike features which include: (i) its absolute size, (ii) position of the centromere, (iii) size of the relative chromosomes, (iv) basic number, and (v) number and position of the satellite. The variation in symmetry is normally associated with chromatin loss, and these two trends (symmetry vs asymmetry) are not essentially correlated with each other (Stebbins 1971). Taking this as lead, Ye et al. (2014) analyzed karyotype of *R. palmatum* which exhibited the karyotype asymmetry index (AsK%; the ratio of the sum of the long arms of a haploid set of chromosomes to the total length of the haploid set of chromosomes × 100) of 55.39% indicating the species to be primitive. Further, and as ascertained by its association with the Stebbins 1A category, karyotype of *R. palmatum* was found to be symmetric like that of *R. tanguticum* (Hu et al. 2007a). Nonetheless, the literature suggests that these two species of *Rheum* are closely related and can hardly be distinguished in the field, though the degree of leaf blade dissection and the shape of its lobes comes to the rescue—*R. palmatum* has lobed leaf blades with narrowly triangular lobes, whereas the leaf blades of *R. tanguticum* are dissected with the lobed parts lanceolate, narrow, and triangular (Bao and Grabovskaya Borodina, 2003)—but again fails to some extent when you come across the transitional morphs between these two species in the field. Almost similar morphological attributes at chromosome level and minimal differences at molecular level (Chen et al. 2008; Wang et al. 2012c) suggests that the two species have a common ancestor, and may even be a single species, i.e., *R. palmatum*. However, more studies (with increased population number and size) are required in light of the current scientific adventures to further examine these species and to validate the inferences drawn from the investigations discussed here. In a related study, Yanping et al. (2011) reported "B" chromosomes in wild populations of *R. tanguticum* as the first report of occurrence of these non-essential supernumerary/accessory chromosomes in genus *Rheum*. The karyotype analysis was done on germinated seedlings of root tips, and the species was found to be diploid (2n = 22; Stebbins 1A type) with a karyotype formula of 2n = 22 = 22m (or 2n = 22

= 2sm + 20m). The number of beta chromosomes observed at metaphase was found to range from one to seven, and these varied not only between individuals, but among different cells of the same root tip as well. The observations (of these extra chromosomes) made in this study confirm that the species of *R. tanguticum* outcross as was earlier suggested by Jones who said that "B" chromosomes are limited to the out-breeders probably because selfing may not assist in the spread and accumulation in populations by mitotic drive (Jones 1995). Another study has reported variation in meiotic chromosome behaviors, viz. lagging and bridging (in addition to that in flower and seed morphologies), in micro-propagated *R. rhaponticum* which was further correlated with the possible role of somaclonal variations in the *in vitro* callus-raised cultures (in contrast to crown-derived plants which showed normal meiotic behavior) of this plant species (Zhao et al. 2008). Nonetheless, few studies have been carried out on various cytological aspects of this important (medicinal) herb which therefore needs to be reassessed on cytogenetic grounds owing to the utility of cytology/karyology in taxonomy and systematics.

6.2 GENETICS

Genus *Rheum* is a highly diversified genus with the mountainous and desert regions of the QTP mainly forming the center of its origin and distribution. As discussed in Chapter 2, the infrageneric classification of rhubarb is primarily based on the morphological characters, and until now, eight sections have been established and acknowledged based on six different types of pollen grains within this genus (Li 1998). This genus symbolizes a virtuous case of the extensive diversification in which the divergent forces largely remain unknown, though some presumptions of convergent evolution cannot be neglected. Pertinently, among the eight recognized sects, the Sect. Globulosa (with globular inflorescence) and Sect. Nobilia (with semi-translucent bracts) are supposed to have an unclear systematic position. Therefore, various attempts have been made wherein state-of-the-art molecular techniques ranging from simple DNA markers [internal transcribed spacer (ITS), maturase K (matK), simple sequence repeats (SSR), inter-simple sequence repeats (ISSR), randomly amplified polymorphic DNA (RAPD), amplified fragment length polymorphism (AFLP), and restriction fragment length polymorphism (RFLP), etc.] to genome and plastome studies have been employed to understand the phylogeny and (growing) genetic diversity of this important genus. This section focuses on the utility of molecular markers in understanding the genetic intricacies of rhubarb at population and species (including vegetable-type cultivars) levels, whereas the genome/transcriptome/plastome-based studies are dealt with in Section 6.3 of this chapter. The objectives of these studies that are briefly discussed below were mainly aimed

1. to redesign the evolutionary relatedness among different species of *Rheum* while examining the history of its diversification;

Molecular Aspects

2. to assess probable relations of some abstruse sections with a unique morphology;
3. to determine possible incidence(s) of the radiative speciation with low genetic divergence within the whole genus vis-à-vis the latest far-reaching and significant uplifts of the QTP;
4. to trace the evolution of certain adaptive traits within the genus, viz. decumbent/caulescent and "glasshouse-like" body plan;
5. to examine the level and pattern of genetic variability and its partitioning within and between the natural populations of different species of *Rheum*;
6. to identify, if any, the specific genetic patterns linked to the geographical distribution of a particular species;
7. to determine whether anthropogenic activities have affected the genetic structure of *Rheum* species;
8. to devise (advised by most of these studies), in light of evaluated genetic diversity, the measures to be adopted to breed/propagate and conserve the better genetic stocks of different species of genus *Rheum*; and finally
9. to define the role and extent of the events of hybridization in diversifying genus *Rheum* in the QTP while employing ITS (existence in multiple copies) in incongruence with cpt DNA.

6.2.1 Rhubarb in the Wild

Polygonaceae are believed to have existed nearly as far back as the Paleocene era with molecular clock hypothesis projecting the divergence of *Rheum* (and its sister groups) dating back to the Miocene. The *Rheum* species are anthophilous and self-incompatible with winged fruits (trigonous achenes) that assist them for wind dispersal. Over this period, various species of *Rheum* are supposed to have accumulated sizeable genetic differences mainly determined by the mating systems in them, like in other plant species (Wang et al. 2012b). In fact, genetic diversity is a very crucial factor for a species to uphold its evolutionary ability to survive environmental perturbations. The loss of this potential genetic diversity is usually related to a compromised fitness of an individual plant species. Indeed, maintaining the genetic variation within and between populations has remained an important factor of conservation for endangered species, because the long-term survival of such a species is firmly reliant on the maintenance of adequate genetic variation to assist adaptations to the changing environmental factors (Cole 2003). To measure genetic diversity and population structure of a particular species at the molecular level, a range of DNA marker techniques are open and very useful. Simple sequence repeats (SSRs) are tandem repeated units of nucleotides which are found copiously in both prokaryotic and eukaryotic genomes. In fact, these repeats are universally dispersed in their genomes' protein-coding and non-coding regions (Tóth et al. 2000). Further, these markers seem to be hypervariable, and, for their codominance and higher reproducibility in such studies, they are mainly regarded as superior to the normally used RAPD, ISSR, and AFLP markers (Wang and Szmidt 2001). Also, many investigations have demonstrated SSR markers to be very effective in determining the genetic variability within and between populations of different plant species (Torres-Díaz et al. 2007).

Earlier, Wang et al. (2005) made a great effort to establish the phylogeny of rhubarb based on sequence analyses of the chloroplast trnL-F region (contains two non-coding cpt DNA sequences, viz. trnL intron and the trnL/trnF intergenic spacer) of 26 species. These species belonged to seven out of eight sections of *Rheum* L. [leaving the Sect. orbicularia (*R. tataricum*)] which nearly covered the whole range of morphological characteristics of this genus. Owing to the fast evolutionary rate and variability as well as the ability to assess the time of divergence for a species within a given lineage, trnL-F region has a wide scope in studies devoted to determine the evolutionary relatedness at intra-/inter-species levels. The study suggests that the rich ecological and geological diversity triggered by the latest extensive uplifts of the QTP might have endorsed speedy speciation patterns (gradual or rapid) in small and isolated populations, in addition to permitting the fixation of exclusive morphological characters in rhubarb. Nonetheless, recognizing the type of speciation, allopatric, or hybridization, which added most to the diversity of these genera, was demarcated as an interesting issue for further research in this area.

Taking the above commentary as its basis, Zhang et al. (2008) in a study focused on the official species of rhubarb (*R. tanguticum*, *R. palmatum*, and *R. officinale*). While aiming to characterize their genetic diversity suggestive of designing suitable conservation measures, ten microsatellite loci (AC/TG/CCA) were primarily developed for *R. tanguticum*. The development of these loci could serve as a valuable tool to ascertain the genetic structure of wild populations of official rhubarb and its allied conservation strategies. In another study by the same research group, and as an extension of the previous work, SSR markers were used to evaluate the extent and partitioning of genetic diversity between the different populations of *R. tanguticum*. This study provides a sumptuous and treasured baseline data on population genetics while encouraging the efforts for both *in situ* and *ex situ* measures to allow the events of gene flow—a central micro-evolutionary force that emphasizes genetic differentiation among populations—and to preserve the gene pools of identified quality genetic groups, respectively (Chen et al. 2009). Nevertheless, it is very time-consuming and expensive to develop specific SSR primers for a targeted species because common SSR primers for all plant species are found to be rare. Therefore, and in comparison to SSR markers, the PCR-amplification-based ISSR markers have been broadly used for the population genetic studies of different plant species. The ISSR microsatellite primers, of whose repeat motifs are abundantly dispersed throughout the genomes, anchor at 3′ or 5′ end and result in amplification of the sequences between the two microsatellite loci. Moreover, the longer sequence and the greater annealing temperature (*Ta*) of ISSR markers make them more reliable with reproducible results than RAPD and other techniques. Besides being free from the need of getting genomic sequence information, ISSR markers are technically simpler and more cost-effective in genetic diversity studies of plants compared to the SSR, AFLP, and RFLP

Molecular Aspects

molecular tools. While taking advantage of the utility of these ISSR markers, and in contrast to the above studies, Hu et al. (2010) evaluated the genetic diversity of the wild populations of the same plant (as studied by above authors), viz. *R. tanguticum* from the Qinghai Province of China. The authors found a substantial correlation between the genetic and geographical distances, wherein the former exhibited a positive correlation with annual mean precipitation and altitude but displayed a negative association with annual mean temperature and latitude. This association probably explains the natural distribution of this important medicinal herb mainly in alpine cold areas. Further, in light of a relatively higher level of genetic diversity in *R. tanguticum* compared to some other Polygonaceous members, the study also proposes formulation of proper conservation (both *in situ* and *ex situ*) studies to maintain its genetic stocks. Owing to the potential of ISSR markers, termed as a dominant technique to evaluate genetic diversity between interrelated species and to spot resemblances among and within species, the official rhubarb species, also treated as important, but endangered and endemic to China, have been subjected to the assessment of genetic diversity by various other groups using ISSR (Wang et al. 2012a, 2012b).

Nevertheless, a recent study by Zhang et al. (2014) claims to have determined the genetic diversity of the similar plant *R. tanguticum* (da huang) on a larger scale which is in contrast to the earlier investigations on this plant species and is also supposed to have overcome the limitations of previous related studies. While anticipating the need for large sample volume and use of new molecular markers, these authors assessed the real state of genetic structure and divergence in Chinese da huang using the matK gene. Positioned within the trnK cpt gene, matK (1.5 kb) is known to be a powerful tool with a higher substitution rate, no interference of heterozygosity, and a speedy evolutionary rate with noted significance in examining the intra-/inter-genus genetic diversity. The study seems to have clarified the genetic structure well and its divergence in da huang from Sichuan, Gansu, and Qinghai provinces of China by employing a new molecular tool, compared to the traditional markers.

The episodes of adaptive radiation form an important source for morphological advancements, ecological diversity, and above all the appearance of new species (Gavrilets and Losos 2009; Givnish et al. 2009). Even though, such incidences are common on aquatic (oceanic) islands (Grant 1999; Baldwin and Sanderson 1998), but may also take place on continental masses if geological/climatic disturbances result in the generation of fresh ecological niches (Arakaki et al. 2011; Richardson et al. 2001; Hughes and Eastwood 2006). One such radiation-linked event is the uplift of the QTP in China (Wan et al. 2014; Wang et al. 2009; Xu et al. 2010; Sun et al. 2012). The diversity in the genus *Rheum* is believed to exist because of the two major radiation events that occurred around 9.9–12 (first) and 5 (second) million years ago. Indeed, huge morphological and ecological differences are visible between various species of *Rheum* from the QTP and its adjoining regions. For instance, some adaptive traits like "decumbent habit" and translucent bract-associated "glasshouse-like" body plan are thought to have evolved many times in parallel over the course of evolution (Zhang et al. 2010; Liu et al. 2013b). With a notion that examining large a number of DNA fragments offers a clear and better resolution of phylogeny between related species (Rowe et al. 2011; Murphy et al. 2001), Sun et al. (2012) attempted to reconstruct the evolutionary relatedness and evaluate the history of divergence of various species of *Rheum* mostly collected from the QTP and nearby areas. While sequencing eight cpt DNA fragments— five coding genes, viz. rbcL, ndhF, psaA, accD, and matK, and three non-coding DNA regions, viz. rbcL–accD, trnK intron, and the trnL-trnF intergenic spacer— of 47 species (34 *Rheum* species and 13 species from related genera), the study further aimed to test the hypothesis of whether or not the morphological traits allied to plant "body plans" (decumbent/caulescent and glasshouse-like body plan) have evolved in parallel in the genus *Rheum*. The investigation clearly confirms the occurrence of rapid radiation in *Rheum* that might have been elicited by the widespread uplifts of the QTP thereby emphasizing the latter's importance. The study further ratifies that during the diversification history of *Rheum*, the two body plan traits have experienced episode(s) of parallel evolution.

It is thus clear that incidence(s) of adaptive radiation result in the rise of characters due to parallel evolution at multiple times. Nonetheless, transmission of traits through hybridization offers an alternate means for these and other characters which appear in seemingly non-sister lineages. Indeed, the sign(s) of hybridization events is quite visible in the disagreements between phylogenies obtained when we use different molecular markers to circumscribe a genus/species, and even from the existence of two differing types of a specific nuclear marker like ITS in a single individual. In this perspective, the above exercise of examining the history of diversification of genus *Rheum* was further extended by the same research group in a study conducted in 2014 wherein they analyzed ITS sequences of thirty species of *Rheum* representing seven sections of this genus to elucidate the possible role of hybridization in its diversity (Wan et al. 2014). In addition to backing the occurrence of at least a single polyploidization event (Hu et al. 2007b; Ruirui et al. 2010), the investigation suggests that extensive hybridization has occurred among the QTP species of genus *Rheum*, though further exploration cannot be ignored. The evidence of this study was based on (i) multiple ITS copies—seven cases of *Rheum* species which contained two divergent ITS versions that ended in unlike phylogenetic positions, in each case signifying hybridization; and (ii) inconsonance between ITS and cpt DNA phylogeny—phylogenetic trees reconstructed based on nrDNA ITS matrix and cpt DNA matrix concerning the position of *R. globulosum* and *R. lhasaense* exposed two more events of hybridization taking the total cases to nine.

6.2.2 Rhubarb Cultivars

As discussed in the beginning of this book in Chapters 1 and 2, rhubarb, with a history dating back to nearly 3000 BC in China forms one of the oldest plants cultivated by humans for medical and later vegetable purposes. In the beginning, the herb was grown exclusively for the medicinal attributes chiefly associated with its roots and became familiarized because of this from Siberia into England

Molecular Aspects

nearly 400 years ago (Cultivation of Rhubarb 1945). In Europe, rhubarb cultivation is believed to have started in the 16th century (Turner 1938) and was grown in Germany and England in the mid-17th century, from where it also spread to the Scandinavian countries (Persson et al. 2000). Later, in the 18th century, a growing interest was seen in the palatable properties of rhubarb petioles (leaf stalks) which made a case for its consumption in vegetable form in the following years. Three different species of readily hybridizing Rheum with edible stalks, viz. *R. undulatum* (= *R. rhabarbarum* according to Englund, 1983), *R. rhaponticum*, and *R. hybridum*, were cultivated as vegetable crops in England and later also in Scandinavia (Turner 1938). Nevertheless, owing to the high drug quality found in Asian rhubarb, European rhubarb cultivation declined in the mid-20th century (Hintze, 1951). The first species of *Rheum* initially planted for medicinal properties was probably *R. rhaponticum* L. Indeed, and although surrounded by some controversies about its possible origin (Stanev 1984), the latter is the only known species native to Europe (Tanhuanpää et al. 2019). Presently, the species is known to exist as an endangered relict plant growing in the Bulgarian mountains (Libert and Englund 1989). Additionally, the plantings of rhubarb for culinary purposes became prevalent only in the 19th century (Cultivation of Rhubarb 1945), and the growers began to develop culinary *R. rhabarbarum* cultivars chiefly in England, which include the still common cultivars, viz. "Victoria" (syn. "Queen Victoria" launched in 1837) and "Prince Albert" (launched in 1840) popularly introduced by Joseph Myatt (Turner 1938). The growth and progress of culinary rhubarbs (*R.* × *hybridum* Murray, synonym *R. V rhabarbarum* L.), generally considered a tetraploid, 2n – 44 (Chin and Youngken 1947), began in England by choosing open-pollinated seeds from *R. undulatum*, *R. rhaponticum*, and *R. hybridum*. To breed new cultivars, *R. hybridum* (anonymous origin, found from 1771 to 1779) was often used as a parent. Moreover, to ensure the maintenance of genotype identity, the cultivars of rhubarb are usually propagated through micro-propagation (Walkey and Matthews 1979) or asexually through crown divisions (Zandstra and DE 1982). Even though propagation through seeds is generally not encouraged as the plants obtained from seeds are not considered to be true to type, it still shows its existence in one way or the other and results in the generation of cultivars with different genotypes (and possibly phenotypes as well) under a common name (Persson et al. 2000). Indeed, different cultivars with similar genotypes are also available; for example, the cultivar "early red" is known by many different names. To supplement this chaos, disorganized crossings and an absence of proper pedigrees from the preliminary pollinations has made it almost impossible to reach to the roots and determine the real origin of today's rhubarb (Turner 1938). In nutshell, the cultivar identification in rhubarb has remained an unsolved and unreachable riddle wrapped in paradox.

The cultivar identification of culinary rhubarb is primarily based on morphological characteristics. In 1999, Rumpunen and Henriksen made an attempt to examine the genetic relatedness/variation between 71 rhubarb cultivars collected from various sources in Denmark from the 1950s (Rumpunen and Henriksen 1999). While evaluating different horticultural traits, primarily the low-oxalate genotypes with better market value, these authors obtained good results of variability by assessing various morphological and biochemical characters (leaf size, leaf vein color, petiole color, petiole width, flowering time, water-soluble oxalate content, and total oxalate content, etc.) aimed to characterize a large collection of rhubarb cultivars. Nonetheless, the utility of such an approach seems narrow as it is sometimes difficult to measure many morphological factors besides the seasonal variations induced due to different environmental factors. For obvious reasons, this remains true for the biochemical characteristics as well. It is believed that these constraints do not act as barriers when suitable genetic markers are employed in such investigations. Therefore, an alternative strategy (preferably based on potential molecular markers) is needed to identify a cultivar, assess its genetic relatedness, and to reach to the obscure origin of the vegetable-type rhubarb where different *Rheum* species seem to be in the running as parents of present-day culinary rhubarb. Notably, just after a single year of the above work, Persson et al. used an integrative approach of evaluating the genetic relatedness/ diversity of rhubarb cultivars, wherein 12 morphological traits and 47 RAPD molecular markers were employed to distinguish 12 culinary rhubarb cultivars (Persson et al. 2000). This study provides an indication of the relatedness and probable common origin of the studied cultivars, and also highlights the utility of the RAPD technique to investigate various other species of *Rheum*. However, Persson could not see considerable correlation between the analyses done on morphological and molecular grounds. Therefore, we must be cautious in generalizing the morphological observations to the molecular inferences.

Later, while modifying the morphological descriptions from previous studies (Persson et al. 2000; Rumpunen and Henriksen 1999), Pantoja and Kuhl (Pantoja and Kuhl 2010) made the first endeavor to determine the genetic diversity in *Rheum* cultivars in the USDA, ARS Rheum collection in Palmer, Alaska. The information generated was loaded in the USDA/ARS, Germplasm Resources Information Network system and made available to the public. The authors present the results of morphological data based on 15 characters of leaf, petiole, and crown of 36 cultivars and two species accessions, viz. *R. palmatum* Rubra and Rheum UK Lot 540533 of rhubarb, collected during 2005 and 2007. The study highlights the variation in the morphological attributes of chosen cultivars/species of *Rheum* in two different (and non-succeeding) years while stressing the need for proper refinement of the selected morphological descriptors. It also suggests making inferences from the results obtained in a single year rather than using data from many years to actually discriminate cultivars with better resolution. In another study, nearly a decade after Persson's work, but a couple of years before the above work of Pantoja and Kuhl (2010), Kuhl and DeBoer (2008) investigated the genetic relatedness/variability in rhubarb cultivars using AFLP as the molecular tool. Four putative species and 37 cultivars of *Rheum* were selected for DNA fingerprinting wherein ten MseI and EcoRI primer combinations were evaluated for about 1400 polymorphisms scored. Being the first report of its kind on culinary rhubarb to assess its genetic diversity, appreciable results were obtained that satisfactorily discriminated the selected cultivars/species on genetic grounds.

Molecular Aspects

With the passage of time, like in wild populations of rhubarb as discussed above, researchers began to evaluate the utility of other molecular tools, viz. ISSR and SSR markers, and the quite discussed/useful ITS regions to assess the genetic relatedness/diversity in the vegetable-type rhubarb cultivars. To examine whether cultivation practices affect the genetics of an important Chinese medicinal herb *R. tanguticum*, Hu et al. (2014) investigated the levels and distribution of genetic variation in five cultivated populations of this perennial herb using ISSR markers. The study suggests that a brief history of domestication with no artificial selection might be an effective way to sustain and conserve the gene pools of *R. tanguticum*. This can further help in designing stratagems for effective and resourceful management of the genetic resources of this and other important plant species. In the same year, and like earlier investigators, Gilmore et al. (2014) also perceived that given the occurrence of synonyms and nomenclatural discrepancies in addition to the environmental effects, it is quite challenging to identify and assess the genetic diversity of culinary rhubarb cultivars with mere morphological attributes. Here again, molecular markers come to the rescue as they are reliable for unique genotypes across environments and deliver the genetic fingerprints to support in-cultivar discrimination. With an aim to utilize a more sophisticated and specific molecular tool than ISSR, these authors developed new di-nucleotide-containing SSR markers (97 novel SSR primer pairs) from the short-read DNA sequences to be utilized for fingerprinting the rhubarb collection of the US National Plant Germplasm System (NPGS) administered by USDA/ARS. To fingerprint various accessions of rhubarb, 25 of the newly developed SSRs were found to be effective in distinguishing each accession selected in this study. This exercise stressed the development and utility of SSR markers screened from the short-read sequencing, besides recommending their efficacious applicability in identifying various horticultural crops. Recently, a major effort was made by Tanhuanpää et al. (2019), wherein the genetic diversity of home-garden rhubarbs on a larger scale was determined all across Finland using 30 SSR markers which were eventually reduced to six for some empirical reasons. Overall, the study contained 647 rhubarb samples (home gardens—539, Luke collection—80, nurseries, etc.—27, and one specimen of *R. palmatum* from USDA/ARS) which were chosen primarily on some selected phenotypic characters, viz. red stems, sweet taste, and early ripening time as well as their age (with about 70 years of cultivation history), location, and stimulating cultural history. The study holds vital importance in generating a quality resource base of genetically varied cultivars representing nearly the whole of Finland, the north-east region in particular. The results obtained served a good purpose to update various accessions of the Finnish national rhubarb collection, wherein duplicates were screened out and the ones with ambiguous origin were supplanted with novel polymorphic ones with precise evidence of their history. This, besides being the first such study at country level, is rather a good attempt to maintain (genetic stocks) and utilize to the full extent the most common cultivars grown in Finland.

China has an age-old documented history of herbal medications with satisfactory health benefits which continues to be in use in contemporary times.

With the growth of herbal markets, the practice of adulteration of materials with medical efficacy has also grown, only to become a concern the world over. These herbal medicines are reportedly known to be purposefully substituted by other related or unrelated plant species which hamper their bioactive potential to a greater extent, which goes in parallel with the level of intended adulteration (Li et al. 2011). As mentioned earlier in this book, da huang (officinal Chinese rhubarb), formed from the dried root and rhizome of *R. officinale*, *R. palmatum*, and *R. tanguticum* (Commission 2010), has wide pharmacological attributes associated with it (Takeoka et al. 2013). Besides medical utility, the diverse benefits of officinal rhubarb have resulted in its use in varied food supplements, increasing its market demand and, side-by-side, making it prone to adulterations and regional substitutes of varied nature. The reportedly known adulterants of officinal rhubarb which more often constitute the root and rhizome of "shan da huang" and "tu da huang" which include various species of *Rheum* (*R. australe*, *R. rhabarbarum*, *R. hotaoense*, *R. franzenbachii*, and *R. wittrockii*) and *Rumex* (*Rumex nepalensis*, *R. obtusifolius*, *R. japonicus*, and *R. chalepensis*) Polygonaceous genera, respectively (Zhou et al. 2017). These mischievous acts of deliberate substitutions/adulterations may lead to varied adverse effects, reduced clinical effectiveness, and the mediocre quality of the concerned medicinal products finally endangering the welfare of consumers. Consequently, with an aim of ensuring the good quality of medication with ensured consumer security, a potent, simple, reliable, and ready-to-use method is required for accurate authentication of da huang (and other herbs of Traditional Chinese Medicine, TCM) along with proper screening and identification of the substituents/adulterants. Taking this as background, Zhang et al. (2013a) used the matK gene sequence to identify the genuine *Rheum* species from its adulterants. Nonetheless, as the matK barcode region is based on organellar DNA (with maternal inheritance), an alternated region (like ITS2) is believed to furnish better information in dealing with such cases as it is inherited from both parents (Chase and Fay 2009). Moreover, owing to certain attributes associated with the ITS2 region, viz. simple, short, and universal with high inter-specific divergence, it has been generally accepted as a common barcode for plant (including herbal medicinal materials) identification and for phylogenetic reconstructions at both the genus and species levels (Yao et al. 2010; Li et al. 2011). While exploiting this property of the ITS2 region, Zhou et al. (2017) analyzed these sequences from officinal source plants sampled from its entire distributional range, as well as a majority of source plant adulterants from 11 provinces. The results obtained from this voluminous investigation suggested that the inter-specific variation between da huang and its adulterants was evidently high compared to the minimal intra-specific variations within the officinal rhubarb species. With a goal of establishing a DNA barcode for valid identification of da huang from its substituents/adulterants for overall consumer safety and its trade in Chinese herbal markets, the authors propose the ITS2 region as a reliable tool for the screening of official rhubarb so that may help in its quality control and therapeutic applications.

Molecular Aspects 119

6.3 BIOTECHNOLOGICAL INTERVENTIONS

The plants serve as natural hotspots for numerous drugs or their precursors used in the current pharmacopoeias as they produce an amazing diversity of low-molecular-weight compounds. Although the structures of around 50,000 have been elucidated to date, there are probably hundreds of thousands of such compounds, with approximately 4000 new ones being discovered every year (Bahadur et al. 2015). The largest group of plant secondary metabolites consists of isoprenoids, comprising more than one-third of all known compounds. The second largest group is formed by alkaloids, which comprise many drugs and poisons. Since the early days of mankind, plants with these secondary metabolites have been used by humans to treat infections, health disorders, and illness. Moreover, and as discussed in the previous chapters, the medicinal plant-based drugs have the added advantage of being simple, effective, and offering a broad spectrum of activity with well-documented prophylactic or curative actions (Verma et al. 2012). Many higher plants are major sources of useful secondary metabolites which are used in the pharmaceutical, agrochemical, flavor, and aroma industries. The World Health Organization (WHO) has estimated that the current demand for medicinal plants is approximately US$14 billion per year, and the demand for medicinal plant-based raw materials is growing at the rate of 15–25% annually. Indeed, according to a WHO estimate, the demand for medicinal plants is likely to increase to more than US$5 trillion in 2050. In India, the medicinal plant-related trade is estimated to be approximately US$1 billion per year (Kala et al. 2006). The search for new plant-derived chemicals should thus be a priority in current and future efforts toward sustainable conservation and rational utilization of biodiversity. Indeed, the metabolomes of medicinal plants serve as a valuable natural resource for pharmaceuticals and provide a platform for renewed attention from both practical and scientific viewpoints for the evidence-based development of new phytotherapeutics and nutraceuticals (Pandith et al. 2014).

It is evident that plants invest a great deal of resources in synthesizing, accumulating, and sorting such metabolites often produced through complex and highly regulated biosynthetic pathways operating in multiple cellular and subcellular compartments (Lewinsohn and Gijzen 2009). The basic biosynthetic pathways that produce plant secondary products, viz. polyketide, shikimate, mevalonate, and non-mevalonate, appear to have been relatively well defined. However, they are constantly being modified as novel compounds are discovered. Additionally, with new secondary metabolites being identified every day, there come new and interesting biochemical pathways that produce them (Romagni 2009). New technologies, such as functional genomics vis-à-vis metabolite production and accumulation, may help to elucidate the sequence of events that lead to the production of secondary compounds. Metabolic engineering is a powerful tool that can be integrated into the natural product drug discovery and development process, from lead optimization to industrial-scale production of the molecules. Advances in our knowledge of natural product biosynthetic pathways, as well as the metabolic and regulatory networks at the genomic level, have allowed for rational development of new compounds (Pickens et al. 2011). Through metabolic engineering, some natural products that can be obtained only in low quantities, e.g., from some plants and marine organisms, can now be produced in engineered heterologous hosts in quantities that are sufficient for industrial application. Metabolic engineering can be a complementary approach to reduce the time needed to optimize a strain, simplify downstream chemical processing, or produce analogs that would be difficult or expensive to access by chemical methods alone.

Genetic modification of pathways of secondary metabolism producing desirable natural products is an attractive approach in plant biotechnology. Engineering a secondary metabolic pathway aims to either increase or decrease the quantity of a certain compound or group of compounds. To decrease the production of a certain unwanted (group of) compound(s), several approaches are possible (Rastegari et al. 2019). More often, the goal is to increase the production of certain compounds in the normal producing plant species or to transfer (part of) a pathway to other plant species or other (micro) organisms. Also, there is interest in the production of novel compounds not yet produced in nature by plants. To increase the production of a (group of) compound(s), two general approaches have been followed. Firstly, methods have been employed to change the expression of one or a few genes, thereby overcoming specific rate-limiting steps in the pathway, to shut down competitive pathways, and to decrease catabolism of the product of interest. Secondly, attempts have been made to change the expression of regulatory genes that control multiple biosynthesis genes (Verpoorte and Memelink 2002). In the past few years, several secondary metabolism genes have been overexpressed in the original plant or in other plant species. However, in practice, there are several major challenges that need to be overcome, such as gaining enough knowledge of the endogenous pathways to understand the best intervention points, identifying and sourcing the most suitable metabolic genes, expressing those genes in such a way as to produce a functional enzyme in a heterologous background, and finally, achieving the accumulation of target compounds without harming the host plant. By using sophisticated molecular biology tools, the detailed knowledge of the pathway of interest and state-of-the-art gene transfer technology, it may become easy to engineer complex metabolic pathways in plants (Farré et al. 2014). A recent review by Gandhi et al. (2015) has comprehensively compiled the contemporary and varied biotechnological approaches widely used to comprehend and utilize the various aspects of secondary metabolism in medicinal (and aromatic) plants for desired results.

The polyketide pathway operates in major plant families, viz. Rhamnaceae, Fabaceae, and Polygonaceae—rhubarb belongs to the latter (Pandith et al. 2020). This pathway, leading to biosynthesis of related key bioactive constituents, is poorly understood as most of the genes operating at various steps of the pathway have seldom been taken to the enzyme level. Moreover, their structural complexity limits their *de novo* synthesis. Thus, naturally occurring and semi-synthetic compounds remain the main sustainable sources for commercial pharmaceutical applications. However, the biosynthesis of specialized metabolites is tightly regulated by developmental, physiological, and environmental factors which limit

Molecular Aspects

their accumulation in natural hosts. Access to such compounds is also often inadequate and a reliance on the production of metabolites from naturally growing plants is not always sustainable. Furthermore, many secondary metabolites are often species-specific in distribution and the plant species in question may be distributed to specific climatic zones as well. Taken together, these issues may limit proper exploitation of plants for large-scale production of economically important compounds. For obvious reasons, a desirable aspect is to improve the levels in native plant species as well as to develop alternative plant or microbial sources/hosts using synthetic biology approaches. In this regard, molecular biotechnological interventions relying on a knowledge of genes and enzymes of a particular biosynthetic pathway offer an attractive approach. Recently, plant biosynthetic pathways have been assembled in engineered microbial systems to produce targeted chemical compounds. For instance, substantial progress has been made using combinatorial biosynthetic approaches to produce high-value bioactive compounds like artemisinic acid (Walter 2014), opiates (Ehrenberg 2015), and aglyconic etoposide (Lau and Sattely 2015) in homologous and/or heterologous hosts. Nonetheless, biogenesis of several important plant secondary metabolites at the level of pathway steps and their regulation is poorly understood, thereby emphasizing the importance of biosynthetic pathway elucidation and detailed exploration of structural and regulatory components. Therefore, understanding the intricacies of biosynthetic pathways holds vital significance for commercial production of specialized metabolites. In fact, the enzymes of specialized metabolic pathways are encoded by small to large gene families, often represented by many homologous members, which makes the pathways more complex. The discovery of putative biosynthetic pathway genes presents a unique challenge owing to the organization of pathway circuitries as complex enzyme networks. Additionally, most of the metabolites which offer a good contribution to the pharmaceutical industry are derived from non-model plants whose genome resources are quite limited (Hall et al. 2013).

The plant genomes contain significant information for comprehending their architecture, and its sequencing will enable the study of relative genomics, as well as serving as a valued source for research on overall plant biology. Therefore, a datamining framework that integrates different "omics" approaches is necessary to efficiently understand, investigate, and further modulate the specialized metabolite pathways in homo- and/or heterologous hosts. Genome sequencing has not only extended our understanding of the blueprints of many plant species but has also revealed the secrets of coding and non-coding genes. Transcriptomics datamining is an efficient way to discover novel pathway genes and/or transcription factors (TFs). In fact, the high-throughput next-generation sequencing (NGS) technologies have revolutionized transcriptomics especially with the advent of RNA-sequencing (RNA-seq) as we are facing an unexpected explosion of released genomes at an unprecedented rate. RNA-seq has now become a more efficient and less expensive tool which is increasingly being used to study the evolutionary origins and ecology of non-model plants (Lee and Hong 2019; Jannesar et al. 2020). Transcriptome analysis not only permits transcript discovery and quantitative analysis of gene expression levels, it also provides unprecedented opportunities to address comparative genomic-level questions, like the processes of speciation or adaptive evolution, between the closely related species of non-model plants.

Additionally, it has been interesting to characterize genes at functional level in order to realize their characteristic roles in plant growth and development. The identity of genes associated with the phenotype of a particular plant involves the customary forward-genetics approaches, viz. positional cloning, etc. (Peters et al. 2003), whereas the methodology of reverse-genetics (T-DNA insertion and tagging, etc.) leads to the evaluation of the functional aspect of a gene (Rosso et al. 2003). Latter (functional genomics), nurtured by genome sequencing, has remarkably allowed significant identification of genes (through NGS) (Parinov and Sundaresan 2000; Morozova and Marra 2008), and their functional characterization by the construction of several gain-of-function (driven by robust promoters, viz. CaMV 35S and/or CaMV 35S enhancers) or loss-of-function (ethane methyl sulfonate mutagenesis or by T-DNA/transposon insertion leading to mutated/ truncated proteins with attenuated/null functions) mutants in different plant species (Teotia et al. 2016). However, these techniques act in a non-specific manner to target a gene. Therefore, to overcome this bottleneck, novel and potent technologies aimed at specific gene targets are needed. Interestingly, both DNA and RNA can be targeted to induce loss-of-function of a specific gene. Various such technologies to target DNA (a specific gene in the genome) have been developed which include zinc-finger nucleases (ZFNs) (Urnov et al. 2010), transcription activator-like effector nucleases (TALENs) (Zhang et al. 2013b), meganucleases (Marcaida et al. 2008), and the clustered regularly interspaced short palindromic repeats (CRISPR)/CRISPR-associated nuclease 9 (Cas9) (CRISPR/Cas9) system (Shan et al. 2013). The latter is an RNA-based DNA cleavage technology, making its application as simple as RNAi but more directional, effective, and diverse than traditional methods for creating genetic mutants (Xie and Yang 2013). Indeed, the techniques of genome editing as guided by the appearance of these sequence-specific nucleases has revolutionized the research on different aspects of basic and applied biology. Following the emergence of CRISPR–Cas9, genome editing became a commonly used technique to characterize gene function and to refine the desired traits. Importantly, the newly established Cas9 variants, innovative RNA-guided nucleases, and base-editing systems, as well as the DNA-free CRISPR–Cas9 delivery methods, now provide great opportunities for plant genome engineering (Yin et al. 2017).

Now coming to the wonder drug, rhubarb; somehow, the plant has received the least attention at the molecular level with some meager reports of utilizing molecular tools primarily to understand its phylogeny to resolve the taxonomic issues within the genus and to determine the ecological functions vis-à-vis alpine adaptations of the species. Indeed, the NCBI database shows a mere ten records of transcriptome sequencing of some of its species, viz. *R. palmatum*, *R. officinale*, *R. rhabarbarum*, *R. rhomboideum*, and *R. nobile*, that too have no or limited public access in unpublished and/or published format (https://www .ncbi.nlm.nih.gov/sra/?term=rheum%20transcriptome&utm_source=gquery

Molecular Aspects 123

&utm_medium=search), and chloroplast sequencing of *R. alexandrae* and *R. tanguticum* (https://www.ncbi.nlm.nih.gov/bioproject/?term=rheum%20chloroplast%20genome&utm_source=gquery&utm_medium=search)—site accessed on February 6, 2020. Moreover, some recent investigations have obtained the plastome sequences of *R. rhabarbarum* (Gao et al. 2019), *R. nobile* (Chen and Li 2020), and *R. pumilum* (Li et al. 2020) without any significant inferences made/ revealed except confirming the phylogenetic position of these species within the Polygonaceae family.

For decades, a variety of phenotypes and physiological machinery adapted by the organisms of arid and alpine habitats to cope with abiotic stressors has remained of topical interest in evolutionary biology (Shao et al. 2007). Pertinently, the molecular basis of natural phenotypic and/or physiological discrepancies associated with plants growing in arid and alpine habitats has become easy to handle with the emergence of ecological genomics (Wright 2004). The peculiar phenotypes of "snowball" and "glasshouse" plants normally observed in alpine environments are presumed to have evolved vis-à-vis the abiotic ecological perturbations including low atmospheric temperature/pressure, high light intensity and robust winds (Tsukaya and Tsuge 2001). *R. nobile*, with specialized "glasshouse" morphology [upper leaves transformed into big luminous creamy bracts covering the whole inflorescence axis to maintain warmth (Terashima et al. 1993) and to filter UV radiation (Omori et al. 2000) to safeguard customary sexual reproduction while escaping normal foliar functions], documented in about ten families (Ohba 1988) including Polygonaceae, are endemic to the alpine zones of the eastern Himalayas. Many investigators have considered it as a model plant to study the "glasshouse" pattern of other such species (Omori and Ohba 2003). With an aim to mainly test the hypothesis that bracts take the least part in photosynthesis, Zhang et al. (2010) investigated this "glasshouse" plant to understand (i) whether normal leaves differ from bracts in foliar structure and photosynthetic attributes, (ii) whether there is variability in genes expressed in the two types of leaves, and (iii) whether the observed changes are related to the translucent bract phenotype and ecological adaptation to alpine/arid habitats. In a nutshell, the study provided evidence on anatomical, physiological, and genetic (cDNA-AFLP analyses) grounds for the functional (physiological) disparity between bracts and the normal leaves in this "glasshouse" species, important in its adaptation to alpine environments. Later, in a related and advanced study, Wang et al. (2014) performed RNA-seq analyses of the same plant, *R. nobile* from the QTP, China, to demonstrate that *de novo* transcriptome sequencing offers an easy, rapid, and cost-effective method to comprehend the relationship between gene expression and complex phenotype evolution at the genomic level. Sequencing of the translucent bract (upper) and rosulate leaf (lower) helped the authors to identify candidate genes (nine flavonoid biosynthesis genes up-regulated in the bract and the mismatch repair related genes) probably involved in alpine adaptation of the large translucent cream-colored bracts in the "glasshouse" plant(s). It further demonstrated the variation in patterns of gene expression essential for the adaptive and intricate evolution of the bracts phenotype. Again, the luminous bracts were found to have derailed from normal-behaving leaves to develop into specialized translucent foliar structures promoting the development of fertile pollen grains. Although the study provides novel insights into bract-associated gene expression profiles, further investigations are required to draw a clear illustration of phenotypic variations vis-à-vis regulatory changes at genetic level. In another study—a comparative investigation to that of Zhang et al. (2010)—Liu et al. (2013a) also tried to understand the adaptive mechanisms of "glasshouse" species to alpine environments while focusing on the only two "glasshouse" species of *Rheum*, viz. *R. alexandrae* and *R. nobile*. The questions which these authors addressed are: (i) Is there any relation between the variation in chloroplast ultrastructure and chlorophyll concentration in the translucent bracts of *R. nobile* and *R. alexandrae*? and (ii) do these two "glasshouse" species share certain common candidate genes for the specialized bracts vis-à-vis their adaptive features? A possible existence of common molecular (cDNA-AFLP analyses) basis for parallel evolution of the "glasshouse" plants was deciphered from the mutual differential gene expressions between bracts and normal leaves of these two related species, thereby emphasizing the significance of the natural selection in generating such phenotypes. Nonetheless, more robust and "omics"-based studies are needed to comprehend the parallel evolution of genetic variations vis-à-vis adaptive evolution of these alpine plants with the specialized "glasshouse" morphology.

Environmental pressures of a varied nature have highly diversified the disputed morphological traits of high-altitude-growing congeneric species of *Rheum* which were also found to be inconsistent with the pollen ornamentations within this genus. This ambiguity was resolved to some extent using cpt DNA fragments, though, a proper taxonomy still remained a challenge due to the rarity of appropriate genetic markers (Sun et al. 2012; Wang et al. 2005; Yang et al. 2001). Consequently, precise genetic markers are required to deduce the phylogeny of *Rheum* species and differentiate them from other allied species. A good number of cpt DNA markers have been utilized for the most sophisticated technique of taxonomic circumscription of various species, DNA barcoding, owing to its highly conserved nature in terms of gene structure and composition compared to the nuclear and mitochondrial genomes (Asaf et al. 2017). Nevertheless, the cpt DNA markers in common use face certain bottlenecks in resolving phylogenies of closely related taxa (Dong et al. 2017). What comes to the rescue here is the whole cpt genome vis-à-vis the advent and easy availability of next-generation sequencing technology. In fact, the complete cpt genome with a conserved sequence length of 110–160 kb that far exceeds the length of ordinarily used molecular markers, is considered to be more informative than simple cpt DNA fragments in determining phylogenetic discrepancies as it contains more hotspot regions with single nucleotide polymorphisms besides the insertion/deletions (InDels) (Yao et al. 2019; Zhou et al. 2018; Guo et al. 2017; Saarela et al. 2018; Nguyen et al. 2017). Taking this commentary as background, a recent investigation by Zhou et al. (2020) focused on cpt genomes of 12 plastomes—ten (eight *Rheum* species, viz. *R. tanguticum*, *R. officinale*, *R. acuminatum*, *R. pumilum*, *R. racemiferum*, *R. hotaoense*, *R. franzenbachii*, and *R. przewalskyi*, and *Rumex crispus*

Molecular Aspects 125

and *Oxyria digyna*) newly sequenced plastomes and two previously published plastomes of *R. palmatum* and *R. wittrockii*—to elucidate the complex phylogeny of Polygonaceae. The results obtained in this study could be useful for specific marker development (some mutation hotspot regions could be tested as *Rheum*-specific DNA markers), valid species discrimination, and the inference of phylogenetic relationships in the genus *Rheum*.

Pertinently, in the pool of least molecular interventions in genus *Rheum*, the role of type III plant polyketide synthases (PKSs) was deciphered in the production of anthraquinones and flavonoids, as major bioactive constituents from *R. emodi* (Syn. *R. australe*), an endangered and endemic medicinal herb of immense therapeutic repute from the north-west Himalayas by us in the recent past (Pandith et al. 2016). The study included an understanding of the functional promiscuity of two divergent chalcone synthase (CHS) gene paralogs due to non-synonymous mutations to drive the chemical diversity. Plausibly, the characterization of *R. australe* polyketide synthases seems essential for complementing flavonoid and/or anthraquinone biosynthesis.

In conclusion, the countable cytological studies on some species of genus *Rheum* advocate the existence of a mere two ploidy levels within it, viz. diploid ($2n = 22$) and tetraploid ($2n = 44$). The detailed karyotyping of a few of its species, viz. *R. palmatum* and *R. tanguticum*, etc., have found that the chromosomes belong to the Stebbins 1A category, i.e., karyotype is symmetrical, although the events of polyploidization and hybridization have also been reported in the genus, but with less frequency. Further, studies related to the appearance of the supernumerary chromosomes in *R. tanguticum* and the abnormal meiotic chromosome behavior in micro-propagated *R. rhaponticum* are also available. Nevertheless, owing to the utility of karyology in cytotaxonomy, further studies need to be undertaken to actually circumscribe the complex genus *Rheum*. Moreover, genetic diversity is an important factor for any species to keep up with the environmental perturbations vis-à-vis its evolutionary ability to survive any external changes. Indeed, maintaining diversity triggered by environmental cues is an essential attribute associated with any species in the wild. The related subsection above has comprehensively covered this critical and dynamic factor as has been maintained by different species of rhubarb (some used as vegetable-type cultivars) mainly in the QTP and its adjoining regions as well as in other places where it occurs or is grown, viz. Denmark, Finland, and the USA (USDA/ARS), etc. The long-term subsistence and evolution of a species is contingent on up-keeping adequate genetic variability within and/or among its populations to accommodate new and challenging selection pressures posed by environmental fluctuations. Therefore, detailed population genetic analyses, as suggested in these investigations, necessarily form one of the primary goals of conservation planning which deserves implementation of substantial efforts. Nevertheless, more studies are required to evaluate the genetic diversity of this important (medicinal and endemic) herb, and to actually discriminate its different cultivars available in contemporary markets using more sophisticated, newly emerging, and efficacious molecular tools. Finally, and like cytology, few studies at molecular level have been carried out on the genus *Rheum* as a whole. This chapter gives a detailed account of these interesting studies, chiefly based on various "omic" (genome/transcriptome/plastome) approaches, and with a detailed impetus for the path which can be followed to understand the molecular biology of this plant vis-à-vis the current technology available. In nutshell, there are many bottlenecks which need to be resolved from all fronts, whether it is cytology, genetics, or molecular biology, to understand the behavior (biological) of this economical herb for its sustainable utility and conservation.

REFERENCES

Arakaki M, Christin P-A, Nyffeler R, Lendel A, Eggli U, Ogburn RM, Spriggs E, Moore MJ, Edwards EJ (2011) Contemporaneous and recent radiations of the world's major succulent plant lineages. *Proceedings of the National Academy of Sciences of the United States of America* 108(20):8379–8384.

Asaf S, Khan AL, Khan MA, Imran QM, Kang S-M, Al-Hosni K, Jeong EJ, Lee KE, Lee I-J (2017) Comparative analysis of complete plastid genomes from wild soybean (Glycine soja) and nine other Glycine species. *PLOS ONE* 12(8):e0182281.

Bahadur B, Rajam MV, Sahijram L, Krishnamurthy K (2015) *Plant Biology and Biotechnology: Volume I: Plant Diversity, Organization, Function and Improvement*. Berlin/Heidelberg, Germany: Springer.

Baldwin BG, Sanderson MJ (1998) Age and rate of diversification of the Hawaiian silversword alliance (Compositae). *Proceedings of the National Academy of Sciences of the United States of America* 95(16):9402–9406.

Bao BJ, Grabovskaya-Borodina AE (2003) Rheum. In: *Flora of China* (Li AR and Bao BJ, eds.), Volume 5. Beijing and Missouri Botanical Garden, St. Louis: Science Press, 341–350.

Chase MW, Fay MF (2009) Ecology. Barcoding of plants and fungi. *Science* 325(5941):682–683.

Chen D, Li L, Zhong G, Qin S, Wang C, Yu Z (2008) Study on genetic relationship of official Rheum by SRAP. *Zhongguo zhong Yao za zhi= zhongguo zhongYao zazhi= China Journal of Chinese Materia Medica* 33(20):2309.

Chen F, Wang A, Chen K, Wan D, Liu J (2009) Genetic diversity and population structure of the endangered and medically important Rheum tanguticum (Polygonaceae) revealed by SSR markers. *Biochemical Systematics and Ecology* 37(5):613–621.

Chen Q, Li Y (2020) The complete chloroplast genome of Rheum nobile. *Mitochondrial DNA Part B* 5(2):1519–1520.

Chin T, Youngken H (1947) The cytotaxonomy of Rheum. *American Journal of Botany* 34(8):401–407.

Cole CT (2003) Genetic variation in rare and common plants. *Annual Review of Ecology, Evolution, and Systematics* 34(1):213–237.

Commission CP (2010) *Pharmacopoeia of the People's Republic of China*, Volume 72. Beijing: China Medical Science Press.

Darlington CD, Janaki Ammal E (1945) *Chromosome Atlas of Cultivated Plants*. London: George Allen & Unwin, p.397.

Darlington CD, Wylie AP (1956) *Chromosome Atlas of Flowering Plants*, Volume 2. London: George Alien & Unwin Ltd., p. 519

Dong W, Xu C, Li W, Xie X, Lu Y, Liu Y, Jin X, Suo Z (2017) Phylogenetic resolution in Juglans based on complete chloroplast genomes and nuclear DNA sequences. *Frontiers in Plant Science* 8:1148.

Molecular Aspects

Ehrenberg R (2015) Engineered yeast paves way for home-brew heroin. *Nature* 521(7552):267.

Englund R (1983) *Cultivated Rhubarb and Taxonomic Problems in the Genus Rheum, Especially the Rhapontica Section.* Report. Sweden: Dept. of Systematic Botany, Uppsala University.

Farré G, Blancquaert D, Capell T, Van Der Straeten D, Christou P, Zhu C (2014) Engineering complex metabolic pathways in plants. *Annual Review of Plant Biology* 65:187–223.

Fedorov AA (1969) *Chromosome numbers of flowering plants.* Leningrad: Academy of Sciences of the USSR, VL Komarov Botanical Institute, Nauka.

Gandhi SG, Mahajan V, Bedi YS (2015) Changing trends in biotechnology of secondary metabolism in medicinal and aromatic plants. *Planta* 241(2):303–317.

Gao H, Tan W, Yang T, Tian X (2019) The complete chloroplast genome of Rheum rhabarbarum. *Mitochondrial DNA Part B* 4(1):1965–1966.

Gavrilets S, Losos JB (2009) Adaptive radiation: Contrasting theory with data. *Science* 323(5915):732–737.

Gilmore BS, Bassil NV, Barney DL, Knaus BJ, Hummer KE (2014) Short-read DNA sequencing yields microsatellite markers for Rheum. *Journal of the American Society for Horticultural Science* 139(1):22–29.

Givnish TJ, Millam KC, Mast AR, Paterson TB, Theim TJ, Hipp AL, Henss JM, Smith JF, Wood KR, Sytsma KJ (2009) Origin, adaptive radiation and diversification of the Hawaiian lobeliads (Asterales: Campanulaceae). *Proceedings of the Royal Society of London. Series B* 276(1656):407–416.

Gohil R, Rather G (1986) Cytogenetic studies of some members of Polygonaceae of Kashmir. III Rheum L. *Cytologia* 51(4):693–700.

Grant PR (1999) *Ecology and Evolution of Darwin's Finches.* Princeton, NJ: Princeton University Press.

Gregory WC (1941) Phylogenetic and cytological studies in the Ranunculaceae Juss. *Transactions of the American Philosophical Society* 31(5):443–521.

Guerra M (2012) Cytotaxonomy: The end of childhood. *Plant Biosystems-An International Journal Dealing with all Aspects of Plant Biology* 146(3):703–710.

Guo X, Liu J, Hao G, Zhang L, Mao K, Wang X, Zhang D, Ma T, Hu Q, Al-Shehbaz IA, Koch MA (2017) Plastome phylogeny and early diversification of Brassicaceae. *BMC Genomics* 18(1):176.

Hall DE, Zerbe P, Jancsik S, Quesada AL, Dullat H, Madilao LL, Yuen M, Bohlmann J (2013) Evolution of conifer diterpene synthases: Diterpene resin acid biosynthesis in lodgepole pine and jack pine involves monofunctional and bifunctional diterpene synthases. *Plant Physiology* 161(2):600–616.

Hintze S (1951) Rhubarb. In: *Swedish plant breeding Part II The garden plants The forest plants. Nature and Culture,* Stockholm:389–391 (in Swedish).

Hu Y, Wang L, Xie X, Yang J, Li Y, Zhang H (2010) Genetic diversity of wild populations of Rheum tanguticum endemic to China as revealed by ISSR analysis. *Biochemical Systematics and Ecology* 38(3):264–274.

Hu Y, Xie X, Feng H (2007a) Karyotype Analysis of Rheum tanguticum Chromosome. *Chinese Pharmaceutical Journal-Beijing* 42(4):258.

Hu Y, Xie X, Wang L, Zhang H, Yang J, Li Y (2014) Genetic variation in cultivated Rheum tanguticum populations. *Genetics and Molecular Biology* 37(3): 540–548.

Hu Y, Xie X, Wen Q, Zhao X, Wang L, Li Y (2007b) Studies on Karyotypes of Five Populations of Rheum Tanguticum Polygonaceae. *Acta Botanica Yunnanica* 29:429–433.

Hughes C, Eastwood R (2006) Island radiation on a continental scale: Exceptional rates of plant diversification after uplift of the Andes. *Proceedings of the National Academy of Sciences of the United States of America* 103(27):10334–10339.

Ilnicki T (2014) Plant biosystematics with the help of cytology and cytogenetics. *Caryologia* 67(3):199–208.

Ilnicki T, Szelag Z (2011) Chromosome numbers in Hieracium and Pilosella (Asteraceae) from central and southeastern Europe. *Acta Biologica Cracoviensia. Series Botanica* 53(1):102–110.

Janaki Ammal E (1955) *Chromosome Atlas of Flowering Plants* (Darlington CD and Wylie AP, eds.), London: George Alien & Unwin Ltd.

Jannesar M, Seyedi SM, Jazi MM, Niknam V, Ebrahimzadeh H, Botanga C (2020) A genome-wide identification, characterization and functional analysis of salt-related long non-coding RNAs in non-model plant Pistacia vera L. using transcriptome high throughput sequencing. *Scientific Reports* 10(1):1–23.

Joachimiak A, Kula A, Grabowska-Joachimiak A (1997) *On Heterochromatin in Karyosystematic Studies.* Acta Biol Cracov Ser Bot 39:69–77. Biological Commission of the Polish Academy of Sciences, Cracow, Poland.

Jones RN (1995) B chromosomes in plants. *New Phytologist* 131(4):411–434.

Kala CP, Dhyani PP, Sajwan BS (2006) Developing the medicinal plants sector in northern India: Challenges and opportunities. *Journal of Ethnobiology and Ethnomedicine* 2(1):32 (https://doi.org/10.1186/1746-4269-2-32).

Kuhl JC, DeBoer VL (2008) Genetic diversity of rhubarb cultivars. *Journal of the American Society for Horticultural Science* 133(4):587–592 (https://doi.org/10.21273/JASHS.133.4.587).

Lau W, Sattely ES (2015) Six enzymes from mayapple that complete the biosynthetic pathway to the etoposide aglycone. *Science* 349(6253):1224–1228.

Lee D-J, Hong CP (2019) Transcriptome atlas by long-read RNA sequencing: Contribution to a reference transcriptome. In: *Transcriptome Analysis.* London: Intech Open Ltd, 25–38.

Lewinsohn E, Gijzen M (2009) Phytochemical diversity: The sounds of silent metabolism. *Plant Science* 176(2):161–169.

Li A (1998) *Flora reipublicae popularis Sinicae: Tomus 25 (1). Angiospermae, Dicotyledoneae, Polygonaceae.* Beijing, China: Science Press, 225–237.

Li M, Cao H, But PPH, Shaw PC (2011) Identification of herbal medicinal materials using DNA barcodes. *Journal of Systematics and Evolution* 49(3):271–283.

Li R, Zhang X, Wang J, Zhou D, Wang H, Shi S, Cheng T (2020) Characterization of the complete chloroplast genome of traditional Tibetan herb, *Rheum pumilum* Maxim. (Polygonaceae). *Mitochondrial DNA Part B, Resources* 5(1):133–135 (https://doi.org/10.1080/23802359.2019.1698344).

Libert B (1987) Breeding a low-oxalate rhubarb (Rheum sp. L.). *Journal of Horticultural Science* 62(4):523–529 (https://doi.org/10.1080/14620316.1987.11515816).

Libert B, Englund R (1989) Present distribution and ecology of *Rheum rhaponticum* (Polygonaceae). *Willdenowia* 19:91–98.

Liu B-B, Opgenoorth L, Miehe G, Zhang D-Y, Wan D-S, Zhao C-M, Jia D-R, Liu J-Q (2013a) Molecular bases for parallel evolution of translucent bracts in an alpine "glasshouse" plant *Rheum alexandrae* (Polygonaceae). *Journal of Systematics and Evolution* 51(2):134–141.

Liu BB, Opgenoorth L, Miehe G, Zhang DY, Wan DS, Zhao CM, Jia DR, Liu JQ (2013b) Molecular bases for parallel evolution of translucent bracts in an alpine "glasshouse" plant Rheum alexandrae (Polygonaceae). *Journal of Systematics and Evolution* 51(2):134–141.

Liu J-Q (2004) Uniformity of karyotypes in Ligularia (Asteraceae: Senecioneae), a highly diversified genus of the eastern Qinghai–Tibet Plateau highlands and adjacent areas. *Botanical Journal of the Linnean Society* 144(3):329–342.

Marcaida MJ, Prieto J, Redondo P, Nadra AD, Alibés A, Serrano L, Grizot S, Duchateau P, Pâques F, Blanco FJ, Montoya G (2008) Crystal structure of I-DmoI in complex with its target DNA provides new insights into meganuclease engineering. *Proceedings of the National Academy of Sciences of the United States of America* 105(44):16888–16893.

Morozova O, Marra MA (2008) Applications of next-generation sequencing technologies in functional genomics. *Genomics* 92(5):255–264.

Murphy WJ, Eizirik E, Johnson WE, Zhang YP, Ryder OA, O'Brien SJ (2001) Molecular phylogenetics and the origins of placental mammals. *Nature* 409(6820):614–618.

Nguyen VB, Park H-S, Lee S-C, Lee J, Park JY, Yang T-J (2017) Authentication markers for five major Panax species developed via comparative analysis of complete chloroplast genome sequences. *Journal of Agricultural and Food Chemistry* 65(30):6298–6306.

Nie Z-L, Wen J, Gu Z-J, Boufford DE, Sun H (2005) Polyploidy in the flora of the Hengduan Mountains hotspot, southwestern China. *Annals of the Missouri Botanical Garden* 92:275–306.

Ohba H (1988) The alpine flora of the Nepal Himalayas: An introductory note. In: *Himalayan Plants* (Ohba H and Malla SB, eds.), Volume 1. Tokyo: University of Tokyo Press, 19–29.

Omori Y, Ohba H (2003) Embryology of a Himalayan glasshouse plant, *Rheum nobile* Hook. f. & Thoms.(Polygonaceae). *Journal of Japanese Botany* 78(3):145–151.

Omori Y, Takayama H, Ohba H (2000) Selective light transmittance of translucent bracts in the Himalayan giant glasshouse plant Rheum nobile Hook. f. & Thomson (Polygonaceae). *Botanical Journal of the Linnean Society* 132(1):19–27.

Pandith SA, Dar RA, Lattoo SK, Shah MA, Reshi ZA (2018) Rheum australe, an endangered high-value medicinal herb of North Western Himalayas: A review of its botany, ethnomedical uses, phytochemistry and pharmacology. *Phytochemistry Reviews : Proceedings of the Phytochemical Society of Europe* 17(3):573–609.

Pandith SA, Dhar N, Rana S, Bhat WW, Kushwaha M, Gupta AP, Shah MA, Vishwakarma R, Lattoo SK (2016) Functional promiscuity of two divergent paralogs of type III plant polyketide synthases. *Plant Physiology* 171(4):2599–2619.

Pandith SA, Hussain A, Bhat WW, Dhar N, Qazi AK, Rana S, Razdan S, Wani TA, Shah MA, Bedi Y, Hamid A, Lattoo SK (2014) Evaluation of anthraquinones from Himalayan rhubarb (Rheum emodi Wall. ex Meissn.) as antiproliferative agents. *South African Journal of Botany* 95:1–8.

Pandith SA, Ramazan S, Khan MI, Reshi ZA, Shah MA (2020) Chalcone synthases (CHSs): The symbolic type III polyketide synthases. *Planta* 251(1):15.

Pantoja A, Kuhl JC (2010) Morphologic variation in the USDA/ARS rhubarb germplasm collection. *Plant Genetic Resources* 8(1):35–41.

Parinov S, Sundaresan V (2000) Functional genomics in Arabidopsis: Large-scale insertional mutagenesis complements the genome sequencing project. *Current Opinion in Biotechnology* 11(2):157–161.

Persson H, Rumpunen K, Möllerstedt L (2000) Identification of culinary rhubarb (Rheum spp.) cultivars using morphological characterization and RAPD markers. *The Journal of Horticultural Science and Biotechnology* 75(6):684–689.

Peruzzi L, Eroğlu HE (2013) Karyotype asymmetry: Again, how to measure and what to measure? *Comparative Cytogenetics* 7(1):1.

Peters JL, Cnudde F, Gerats T (2003) Forward genetics and map-based cloning approaches. *Trends in Plant Science* 8(10):484–491.

Pickens LB, Tang Y, Chooi Y-H (2011) Metabolic engineering for the production of natural products. *Annual Review of Chemical and Biomolecular Engineering* 2:211–236.

Rastegari AA, Yadav AN, Yadav N (2019) Genetic manipulation of secondary metabolites producers. In: *New and Future Developments in Microbial Biotechnology and Bioengineering*. Amsterdam, The Netherlands: Elsevier:13–29.

Richardson JE, Pennington RT, Pennington TD, Hollingsworth PM (2001) Rapid diversification of a species-rich genus of Neotropical rain forest trees. *Science* 293(5538):2242–2245.

Romagni J (2009) Biosynthesis of Chemical Signals—De Novo Synthesis and Secondary Metabolites. In: Hardege JD (ed) *Chemical Ecology, Encyclopedia of Life Support Systems*. UNESCO. EOLSS Publishers, United Kingdom.

Rosso MG, Li Y, Strizhov N, Reiss B, Dekker K, Weisshaar B (2003) An Arabidopsis thaliana T-DNA mutagenized population (GABI-Kat) for flanking sequence tag-based reverse genetics. *Plant Molecular Biology* 53(1–2):247–259.

Rowe KC, Aplin KP, Baverstock PR, Moritz C (2011) Recent and rapid speciation with limited morphological disparity in the genus Rattus. *Systematic Biology* 60(2):188–203.

Ruirui L, Wang A, Tian X, Wang D, Liu J (2010) Uniformity of karyotypes in Rheum (Polygonaceae), a species-rich genus in the Qinghai-Tibetan Plateau and adjacent regions. *Caryologia* 63(1):82–90.

Rumpunen K, Henriksen K (1999) Phytochemical and morphological characterization of seventy-one cultivars and selections of culinary rhubarb (Rheum spp.). *The Journal of Horticultural Science and Biotechnology* 74(1):13–18.

Saarela JM, Burke SV, Wysocki WP, Barrett MD, Clark LG, Craine JM, Peterson PM, Soreng RJ, Vorontsova MS, Duvall MR (2018) A 250 plastome phylogeny of the grass family (Poaceae): Topological support under different data partitions. *PeerJ* 6:e4299.

Saggoo MIS, Farooq U (2011) Cytology of Rheum, a vulnerable medicinal plant from Kashmir Himalaya. *Chromosome Botany* 6(2):41–44.

Shan Q, Wang Y, Li J, Zhang Y, Chen K, Liang Z, Zhang K, Liu J, Xi JJ, Qiu J-L, Gao C (2013) Targeted genome modification of crop plants using a CRISPR-Cas system. *Nature Biotechnology* 31(8):686–688.

Shao H-B, Guo Q-J, Chu L-Y, Zhao X-N, Su Z-L, Hu Y-C, Cheng J-F (2007) Understanding molecular mechanism of higher plant plasticity under abiotic stress. *Colloids and Surfaces: Biointerfaces* 54(1):37–45 (https://doi.org/10.1016/j.colsurfb.2006.07 .002).

Siljak-Yakovlev S, Peruzzi L (2012) Cytogenetic characterization of endemics: Past and future. *Plant Biosystems-An International Journal Dealing with all Aspects of Plant Biology* 146(3):694–702.

Stace CA (1991) *Plant Taxonomy and Biosystematics*. Cambridge, England: Cambridge University Press.

Stanev S (1984) Rheum rhaponticum. *Červena Kniga na NR Bălgarija* 90:1-Sofija. Centre for Studies in Architecture and Urbanism (CEAU), the Faculty of Architecture, University of Porto, Portugal.

Stebbins GL (1971) *Chromosomal Evolution in Higher Plants. Chromosomal Evolution in Higher Plants*, Volume 1. London: Edward Arnold Ltd.:8–216.

Sun Y, Wang A, Wan D, Wang Q, Liu J (2012) Rapid radiation of Rheum (Polygonaceae) and parallel evolution of morphological traits. *Molecular Phylogenetics and Evolution* 63(1):150–158.

Takeoka GR, Dao L, Harden L, Pantoja A, Kuhl JC (2013) Antioxidant activity, phenolic and anthocyanin contents of various rhubarb (R heum spp.) varieties. *International Journal of Food Science and Technology* 48(1):172–178.

Molecular Aspects 131

Tanhuanpää P, Suojala-Ahlfors T, Hartikainen M (2019) Genetic diversity of Finnish home garden rhubarbs (Rheum spp.) assessed by simple sequence repeat markers. *Genetic Resources and Crop Evolution* 66(1):17–25.

Taylor HV, Skillman EE (1945) Cultivation of Rhubarb. *Nature* 155(3934):359–359. doi:10.1038/155359b0.

Te Beest M, Le Roux JJ, Richardson DM, Brysting AK, Suda J, Kubešová M, Pyšek P (2012) The more the better? The role of polyploidy in facilitating plant invasions. *Annals of Botany* 109(1):19–45.

Teotia S, Singh D, Tang X, Tang G (2016) Essential RNA-based technologies and their applications in plant functional genomics. *Trends in Biotechnology* 34(2): 106–123.

Terashima I, Masuzawa T, Ohba H (1993) Photosynthetic characteristics of a giant alpine plant, *Rheum nobile Hook. f. et Thoms.* and of some other alpine species measured at 4300 m, in the Eastern Himalaya, Nepal. *Oecologia* 95(2):194–201 (https://doi .org/10.1007/BF00323490).

Torres-Díaz C, Ruiz E, González F, Fuentes G, Cavieres LA (2007) Genetic diversity in Nothofagus alessandrii (Fagaceae), an endangered endemic tree species of the Coastal Maulino Forest of Central Chile. *Annals of Botany* 100(1):75–82.

Tóth G, Gáspári Z, Jurka J (2000) Microsatellites in different eukaryotic genomes: Survey and analysis. *Genome Research* 10(7):967–981.

Tsukaya H, Tsuge T (2001) Morphological adaptation of inflorescences in plants that develop at low temperatures in early spring: The convergent evolution of "downy plants". *Plant Biology* 3(5):536–543.

Turner D (1938) The economic rhubarbs: A historical survey of their cultivation in Britain. *Journal of the Royal Horticultural Society* 63:355–360.

Urnov FD, Rebar EJ, Holmes MC, Zhang HS, Gregory PD (2010) Genome editing with engineered zinc finger nucleases. *Nature Reviews Genetics* 11(9):636–646.

Verma P, Mathur AK, Jain SP, Mathur A (2012) *In vitro* conservation of twenty-three overexploited medicinal plants belonging to the Indian sub-continent. *The Scientific World Journal* (https://doi.org/10.1100/2012/929650).

Verpoorte R, Memelink J (2002) Engineering secondary metabolite production in plants. *Current Opinion in Biotechnology* 13(2):181–187.

Walkey D, Matthews K (1979) Rapid clonal propagation of rhubarb (Rheum rhaponticum L.) from meristem-tips in tissue culture. *Plant Science Letters* 14(3):287–290.

Walter JM (July 20–24, 2014) Engineering yeast to produce artemisinic acid for anti-malarial drugs. In: *Annual Meeting and Exhibition* (July 20–24, 2014). (https://sim .confex.com/sim/2014/webprogram/Paper27488.html).

Wan D, Sun Y, Zhang X, Bai X, Wang J, Wang A, Milne R (2014) Multiple ITS copies reveal extensive hybridization within Rheum (Polygonaceae), a genus that has undergone rapid radiation. *PLOS ONE* 9(2):e90346.

Wang A, Yang M, Liu J (2005) Molecular phylogeny, recent radiation and evolution of gross morphology of the rhubarb genus Rheum (Polygonaceae) inferred from chloroplast DNA trn LF sequences. *Annals of Botany* 96(3):489–498.

Wang L, Abbott RJ, Zheng W, Chen P, Wang Y, Liu J (2009) History and evolution of alpine plants endemic to the Qinghai-Tibetan Plateau: Aconitum gymnandrum (Ranunculaceae). *Molecular Ecology* 18(4):709–721.

Wang L, Zhou H, Han J, Milne RI, Wang M, Liu B (2014) Genome-scale transcriptome analysis of the alpine "glasshouse" plant Rheum nobile (Polygonaceae) with special translucent bracts. *PLOS ONE* 9(10):e111988.

Wang X-M, Hou X-Q, Zhang Y-Q, Yang R, Feng S-F, Li Y, Ren Y (2012a) Genetic diversity of the endemic and medicinally important plant Rheum officinale as revealed by inter-simpe sequence repeat (ISSR) markers. *International Journal of Molecular Sciences* 13(3):3900–3915.

Wang X-R, Szmidt AE (2001) Molecular markers in population genetics of forest trees. *Scandinavian Journal of Forest Research* 16(3):199–220.

Wang X, Yang R, Feng S, Hou X, Zhang Y, Li Y, Ren Y (2012b) Genetic variation in Rheum palmatum and Rheum tanguticum (Polygonaceae), two medicinally and endemic species in China using ISSR markers. *PLOS ONE* 7(12):e51667.

Wang X, Yang R, Feng S, Hou X, Zhang Y, Li Y, Ren Y (2012c) Genetic variation in Rheum palmatum and Rheum tanguticum (Polygonaceae), two medicinally and endemic species in China using ISSR markers. *PLOS ONE* 7(12):e51667.

Wissemann V (2007) Plant evolution by means of hybridization. *Systematics and Biodiversity* 5(3):243–253.

Wright BE (2004) Stress-directed adaptive mutations and evolution. *Molecular Microbiology* 52(3):643–650.

Xie K, Yang Y (2013) RNA-guided genome editing in plants using a CRISPR–Cas system. *Molecular Plant* 6(6):1975–1983.

Xu T, Abbott RJ, Milne RI, Mao K, Du FK, Wu G, Ciren Z, Miehe G, Liu J (2010) Phylogeography and allopatric divergence of cypress species (Cupressus L.) in the Qinghai-Tibetan Plateau and adjacent regions. *BMC Evolutionary Biology* 10(1):194.

Yang M, Zhang D, Zheng J, Liu J (2001) Pollen morphology and its systematic and ecological significance in Rheum (Polygonaceae) from China. *Nordic Journal of Botany* 21(4):411–418.

Yanping H, Wang L, Li YJC (2011) New Occurrence of B Chromosomes in *Rheum tanguticum* Maxim. Ex Balf.(Polygonaceae). Caryologia 64(3):320–324 (https://doi.org/10.1080/00087114.2011.10589798).

Yao G, Jin J-J, Li H-T, Yang J-B, Mandala VS, Croley M, Mostow R, Douglas NA, Chase MW, Christenhusz MJ, Soltis DE, Soltis PS, Smith SA, Brockington SF, Moore MJ, Yi TS, Li DZ (2019) Plastid phylogenomic insights into the evolution of Caryophyllales. *Molecular Phylogenetics and Evolution* 134:74–86.

Yao H, Song J, Liu C, Luo K, Han J, Li Y, Pang X, Xu H, Zhu Y, Xiao P, Chen S (2010) Use of ITS2 region as the universal DNA barcode for plants and animals. *PLOS ONE* 5(10):e13102.

Ye J, Jia Y, Fan K, Sun X, Wang XJG (2014) Research M. *Karyotype Analysis of Rheum palmatum Genetics and Molecular Research* 13(4):9056–9061.

Yin K, Gao C, Qiu J-L (2017) Progress and prospects in plant genome editing. *Nature Plants* 3:17107 (doi: 10.1038/nplants.2017.107).

Zandstra B, DE M (1982) *A Grower's Guide to Rhubarb Production*. East Lansing, MI: Michigan State University. 30(12):6–10.

Zhang D-Y, Chen N, Yang Y-Z, Zhang Q, Liu J-Q (2008) Development of 10 microsatellite loci for Rheum tanguticum (Polygonaceae). *Conservation Genetics* 9(2):475–477.

Zhang D, Liu B, Zhao C, Lu X, Wan D, Ma F, Chen L, Liu J (2010) Ecological functions and differentially expressed transcripts of translucent bracts in an alpine 'glasshouse' plant Rheum nobile (Polygonaceae). *Planta* 231(6):1505–1511.

Zhang X, Liu C, Yan X, Cheng X, Liu J, Wang Q, Liu K, Wei S (2013a) Sequence analysis and identification of a chloroplast matK gene in Rhei rhizoma from different botanical origins. *Yao Xue Xue Bao= Acta Pharmaceutica Sinica* 48(11):1722–1728.

Zhang X, Liu Y, Gu X, Guo Z, Li L, Song X, Liu S, Zang Y, Li Y, Liu C, Wei S (2014) Genetic diversity and population structure of Rheum tanguticum (Dahuang) in China. *Chinese Medicine* 9(1):26.

Molecular Aspects 133

Zhang Y, Zhang F, Li X, Baller JA, Qi Y, Starker CG, Bogdanove AJ, Voytas DF (2013b) Transcription activator-like effector nucleases enable efficient plant genome engineering. *Plant Physiology* 161(1):20–27.

Zhao Y, Zhou Y, Grout BW (2008) Alterations in flower and seed morphologies and meiotic chromosome behaviors of micropropagated rhubarb (Rheum rhaponticum L.)'PC49'. *Scientia Horticulturae* 116(2):162–168.

Zhou T, Wang J, Jia Y, Li W, Xu F, Wang X (2018) Comparative chloroplast genome analyses of species in Gentiana section Cruciata (Gentianaceae) and the development of authentication markers. *International Journal of Molecular Sciences* 19(7):1–15 (http://dx.doi.org/10.3390/ijms19071962).

Zhou T, Zhu H, Wang J, Xu Y, Xu F, Wang X (2020) Complete chloroplast genome sequence determination of Rheum species and comparative chloroplast genomics for the members of Rumiceae. *Plant Cell Reports* 39(6):811–824.

Zhou Y, Du X-L, Zheng X, Huang M, Li Y, Wang X-M (2017) ITS2 barcode for identifying the officinal rhubarb source plants from its adulterants. *Biochemical Systematics and Ecology* 70:177–185.

7 Conservation

7.1 THREAT STATUS

Rhubarb is a well-known medicinal and vegetable perennial herb distributed mainly in central and northern Asia. There are good reports of household use of this herb as well, and it is cooked as a vegetable or served in the form of various recipes/desserts with added sugar. The species of rhubarb, primarily being medicinal with high therapeutic reputation, are among the most highly pursued species for varied purposes, which range from mere consumption to even as a trade item by the people where rhubarb is endemic. People harvest the vegetative part of the plant for use as a vegetable or for other homemade recipes, while the roots/rhizomes are collected, dried, and stored for use in several herbal formulations against many human ailments. Indeed, over recent decades, various species of this sought-after herb have been exposed to diverse natural and anthropogenic threats in their native regions, leading to a rapid decline in the natural habitat. The local pressures include rapid urbanization, industrialization, uncontrolled deforestation, landslides, high tourist numbers exceeding the usual carrying capacity of specific health resorts, construction of roads through areas occupied by natural vegetation, precision illegal extraction, overgrazing, and overexploitation for local use among others (Kabir Dar et al. 2015; Pandith et al. 2018). Nature has selected beautiful, though widely grazed, alpine meadows as the habitat of this Polygonaceous herb where its distribution and population size are harshly limited by grazing herbivores which tend to threaten this economically important medicinal plant (Tali et al. 2015). Moreover, as per the authors' personal observation, certain species of rhubarb (Himalayan/Indian rhubarb) could be spotted with very thin distribution after trekking for long periods in the higher reaches of the north-west Indian Himalayas; the case is completely antagonistic to the observations made earlier by us and by other people working in the area. Certainly, due to the natural and anthropogenic stressors, the plant has made a shift from lower to higher altitudes with a sporadic distribution pattern over the preceding decades (Kabir Dar et al. 2015; Tali et al. 2015). This distressing statistic in nature of the availability of vulnerable species of genus *Rheum* is indeed a major concern for mankind in general, and taxonomists in particular. Owing to the magnitude of diminishing populations of rhubarb in natural environments, current literature reports two species of *Rheum*, *R. australe* and *R. webbianum*, as endangered (Pandith et al. 2016; Rashid et al. 2014). Such a situation definitely calls for urgent attention to arrest and reverse this losing trend by making strategic efforts for conservation and sustainable utilization vis-à-vis the regional economic benefits of this important medicinal herb. Pertinently, for apposite conservation planning, it is imperative to have a detailed knowledge of the biology of these species, their

DOI: 10.1201/9780429340390-7

135

population status, natural and/or anthropogenic factors affecting the size of their populations, and the fitness of individuals within the populations. Moreover, to avoid the harmful impact of exterior threats on these important plant species and to safeguard them from becoming threatened, proper and instantaneous *in situ* and *ex situ* conservation measures need to be employed for their sustainable usage and development. The practice of cultivation shall be molded toward standardization, as apparent from certain topical investigations on *R. australe* (Bano et al. 2017; Sharma and Sharma 2017). This may follow the establishment of breeding zones, free-standing of its natural habitat including growth of progressive agro-techniques and generation of superior plant material to increase its population size in unaccustomed habitats. Additionally, regional farmers/growers can be fortified for commercial and profitable cultivation of rhubarb for varied economic dividends that may also relieve the burden on their wild relatives. Nevertheless, and unfortunately, in the Himalayan context, the misplaced priorities, inadequate sampling, incorrect identification, and inappropriate threat status hints toward the significance of goal-oriented and robust studies that are more important to attribute the real threat status to such species, as per the latest IUCN criteria which will follow the implementation of a required and sophisticated conservation strategy. In fact, a strategy of assessing, documenting, monitoring, and reintroducing this (*Rheum*) and other threatened species of high therapeutic repute holds pivotal significance in promoting community-managed conservation vis-à-vis livelihood improvement of regional mountain communities. In a nutshell, there is dire need to take promising and operative measures to understand and then ease the overall effect of current threats to rhubarb.

7.2 *IN VITRO* PROPAGATION STUDIES AS A CONSERVATION MEASURE

Endemic and threatened plant species are a vital component of plant biodiversity which require immediate human intervention to ensure their long-term survival. The conservation of wild medicinal plants or any other such threatened species can be tackled by scientific techniques as well as social actions. The practice of an *in vitro* micro-propagation/regeneration scheme offers a vibrant advantage over naturally growing individuals as it has the potential to provide an organized production system that guarantees a regular stock with uniform quality and yield. Certainly, in comparison to the long-standing conventional approaches of multiplication of a particular species with some significance, the tissue culture method has proven to be a useful asset for easy production of similar individuals of a particular genotype (Pandith et al. 2018). Indeed, the technique of *in vitro* propagation of important medicinal and aromatic plants is one of the most common and easy methods, which acts as a convenient prelude to the conservation of the flora of significance. Tissue culture is basically the aseptic *in vitro* production of plants from organs, tissues, cells, or even protoplasts in organized and hygienic environmental settings (Bhojwani and Razdan 1986). There is not a single universally applicable method available for culture and regeneration of

Conservation

the vast majority of plant species as it is a delicate and quite variable process. Moreover, the whole plant regeneration from an explant depends upon the processes of totipotency and plasticity of plant cells. The concept of cellular totipotency was advocated by Haberlandt who made pioneering attempts to culture isolated and fully differentiated cells (Haberlandt 1902). However, formal plant tissue culture studies were undertaken only after the identification of auxin in the 1930s and importance of B vitamins for plant growth were realized (Aberg 1961). The discovery of cytokinin as a plant growth regulator and its effectiveness with auxins to direct organ formation was an important hallmark for developing reliable methods for plant regeneration from cultured cells (Skoog and Miller 1957). These studies were coincident with the independent discoveries of the development of somatic embryos by Steward and Reinert (Steward et al. 1958; Reinert 1959). The processes of organogenesis and embryogenesis led to the regeneration of plants under aseptic *in vitro* conditions. In the former, plant regeneration takes place as a result of the meristematic activity of shoot buds, whereas the latter gives rise to somatic embryos from a differentiation event of a single cell or cluster of cells. Such processes of differentiation or morphogenesis can either be indirect via de-differentiated callus or direct through an explant (Chandra and Pental 2003). The preliminary investigation on *in vitro* propagation of the perennial herb R. *rhaponticum* L. was done by Roggemans wherein he noticed initiation of the axillary buds when the plant was grown on MS medium fortified with 1 mg/l of 3-indolebutyric acid (IBA) and 1 mg/l of 6 benzylaminopurine (BAP). IBA, when supplied alone, induced rooting and the rooted plantlets were processed for hardening with about a 70% success rate (Roggemans and Claes 1979). Following the observations made by Roggemans, Walkey worked on the *in vitro* multiplication and regeneration ability of rhubarb (Walkey and Matthews 1979). He observed the production of profuse proliferating units from inoculated meristem-tips on MS medium supplemented with 12.8 mg/l of kinetin; rooting was detected from the explants grown on a phytohormone-less medium. Later on, further studies, though inadequate, were focused on *in vitro* propagation of other species of *Rheum*. Kozak and Salata, while initiating *in vitro* cultures from buds isolated from three-year-old crowns of *R. rhaponticum*, studied the effect of four cytokinins, viz. BA, kinetin, 2iP (isopentenyladenine; with best effect at 12.3 μmol dm^{-3}), and TDZ (thidiazuron), on shoot multiplication of this popular vegetable crop (Kozak and Salata 2011). Moreover, Rayirath et al. explicated the role of ethylene and jasmonic acid in the induction and growth/development of miniature rhizomes in *R. rhabarbarum* under *in vitro* conditions (Rayirath et al. 2011). Rashid et al. obtained a callus from rhizome explants of *R. webbianum* when grown on MS medium supplemented with different auxin concentrations, with the best response seen with 0.5 mg/l 2,4-D. The callus was later shown to differentiate into shoots when the medium was fortified with various concentrations of IAA, IBA, and BAP (Rashid et al. 2014). A detailed report on the *in vitro* multiplication of one of the important species of rhubarb, *R. australe*, is available in our recent communication on the species (Pandith et al. 2018). Further, a handful of additional studies have focused on the *in vitro* propagation of other species of *Rheum*, viz. *R. moorcroftianum* (Maithani 2015), *R. palmatum* (Ishimaru et al. 1990; Kasparova and Siatka 2001a, b; Cui et al. 2008), *R. officinale* (Ji-yong 2010), *R. ribes* (Sepehr and Ghorbanli 2005), *R. rhabarbarum* (Lepse 2007), *R. tanguticum* (Xu et al. 2004), and *R. coreanum* (Mun and Mun 2016), etc. A book chapter on the *in vitro* culture of rhubarb species and sennosides production is also available (Shibata 1993). Nonetheless, and owing to the pharmacological utility of rhubarb species, these reports do not account much for the *in vitro* regeneration/multiplication system of this medicinal herb. But, as a prelude, it provides a platform for advanced empirical studies, imperative for a better and valuable output. In our own laboratory, we have established well-developed and reproducible *in vitro* regeneration systems of two species of *Rheum*, viz. *R. australe* and *R. spiciforme*, on MS medium supplemented with 3% (w/v) sucrose. The medium was supplemented with different combinations and varying concentrations of different phytohormones to observe the events of callus induction, somatic embryogenesis, shoot multiplication, and root initiation (data unpublished).

In conclusion, owing to the dwindling populations of rhubarb in natural habitats due to various kinds of unchecked human interventions, as well as the harsh regional natural conditions, the plant species deserve immediate and effective attention to formulate sustainable conservation measures. Nevertheless, to design a strategy for the conservation and management of such a threatened endemic species, it is necessary to identify their current geographic distribution and obtain information about their abundance, genetic variability, population dynamics, and reproductive ecology (López-Toledo et al. 2011). By all means, productive approaches toward the *ex situ* and *in situ* conservation strategies should be formulated and further implemented to preserve and make sustainable use of this threatened species. Pertinently, owing to the rich and indigenous knowledge acquired through generations by mountain communities, a community-based approach for the management and sustainable use of dependent, but limited, bioresources is a vital approach to enabling both the conservation of biodiversity as well as supporting local livelihoods. This approach could facilitate and enhance the process of making the national- and state-level policies and programs more responsive to linkages between sustainable rural livelihoods and biodiversity conservation, which would go a long way to conserve not only the target species but will also lay a framework that could be followed for other such species which merit attention.

REFERENCES

Aberg B (1961) VitaminS as growth factors in higher plants. In: Ruhland W (ed) *Handbuch der Pflanzenphysiologi XIV. Growth and growth substances.* Berlin: Springer:418–449.

Bano H, Siddique MAA, Gupta RC, Bhat MA, Mir SA (2017) Response of Rheum australe L.(rhubarb),(Polygonaceae) an endangered medicinal plant species of Kashmir Himalaya, to organic-inorganic fertilization and its impact on the active component Rhein. *Journal of Medicinal Plants Research* 11(6):118–128.

Bhojwani SS, Razdan MK (1986) *Plant Tissue Culture: Theory and Practice*, Volume 5. Amsterdam, The Netherlands: Elsevier.

Conservation

Chandra A, Pental D (2003) Regeneration and genetic transformation of grain legumes: An overview. *Current Science* 84(3):381–387.

Cui Y, Liu X, Han J, Wang B, Guo D (2008) Biotransformation of podophyllotoxin by cell suspension culture and root culture of *Rheum palmatum*. Zhongguo Zhong Yao za Zhi = Zhongguo Zhongyao Zazhi = China Journal of Chinese Materia Medica 33(9):989–991.

Farzami MS, Ghorbant M (2005) Formation of Catechin in Callus Cultures and Micropropagation of *Rheum ribes* L. *Pakistan Journal of Biological Sciences* 8(10):1346–1350.

Haberlandt G (1902) Plant cell culture experiment with isollierten. *SB Vienna Ways Science* 111:69–92.

Ishimaru K, Satake M, Shimomura K (1990) Production of (+)-Catechin in Root and Cell Suspension Cultures of *Rheum palmatum* L. *Plant Tissue Culture Letters* 7(3):159–163 (https://doi.org/10.5511/plantbiotechnology1984.7.159).

Ji-yong J (2010) Tissue culture of rhubarb. *Yinshan Academic Journal* (Natural Science Edition) 2.

Kabir Dar A, Siddiqui M, Wahid-ul H, Lone A, Manzoor N, Haji A (2015) Threat status of *Rheum emodi*-A study in selected cis-Himalayan regions of Kashmir valley Jammu & Kashmir India. *Medicinal and Aromatic Plants* 4:183.

Kasparova M, Siatka T (2001a) Effect of chitosan on the production of anthracene derivatives in tissue culture of *Rheum palmatum* L. *Ceska a Slovenska Farmacie: Casopis Ceske Farmaceuticke Spolecnosti a Slovenske Farmaceuticke Spolecnosti* 50(5):249–253.

Kasparova M, Siatka T (2001b) Effect of the biotic elicitor, Candida utilis, on the production of anthracene derivatives in a tissue culture of *Rheum palmatum* L. *Ceska a Slovenska Farmacie : Casopis Ceske Farmaceuticke Spolecnosti a Slovenske Farmaceuticke Spolecnosti* 50(1):41–45. PMID: 11242835.

Kozak D, Salata A (2011) Effect of cytokinins on in vitro multiplication of rhubarb (Rheum rhaponticum L.)'Karpow Lipskiego'shoots and ex vitro acclimatization and growth. *Acta Scientiarum Polonorum: Hortorum Cultus* 10:75–87.

Lepse L (2007) Comparison of in vitro and Traditional Propagation Methods of Rhubarb (Rheum rhabarbarum) according to Morphological Features and Yield. In: *International Symposium on Acclimatization and Establishment of Micropropagated Plants ISHS Acta Horticulturae 812* III(812):265–270.

López-Toledo L, Gonzalez-Salazar C, Burslem DF, Martinez-Ramos M (2011) Conservation assessment of Guaiacum sanctum and Guaiacum coulteri: Historic distribution and future trends in Mexico. *Biotropica* 43(2):246–255.

Maithani U (2015) In-vitro Propagation Studies of Rheum moorcroftianum Royle: A Threatened Medicinal plant from Garhwal Himalaya. *International Journal of Current Microbiology and Applied Sciences* 4(6):596–599.

Mun S-C, Mun G-S (2016) Development of an efficient callus proliferation system for Rheum coreanum Nakai, a rare medicinal plant growing in Democratic People's Republic of Korea. *Saudi Journal of Biological Sciences* 23(4):488–494.

Pandith SA, Dar RA, Lattoo SK, Shah MA, Reshi ZA (2018) Rheum australe, an endangered high-value medicinal herb of North Western Himalayas: A review of its botany, ethnomedical uses, phytochemistry and pharmacology. *Phytochemistry Reviews : Proceedings of the Phytochemical Society of Europe* 17(3):573–609.

Pandith SA, Dhar N, Rana S, Bhat WW, Kushwaha M, Gupta AP, Shah MA, Vishwakarma R, Lattoo SK (2016) Functional promiscuity of two divergent paralogs of Type III plant polyketide synthases from Rheum emodi Wall ex. Meissn. *Plant Physiology* 171(4):2599–2619.

Rashid S, Kaloo ZA, Singh S, Bashir I (2014) Callus induction and shoot regeneration from rhizome explants of Rheum webbianum Royle-a threatened medicinal plant growing in Kashmir Himalaya. *Journal of Scientific and Innovative Research* 3(5):515–518.

Rayirath UP, Lada RR, Caldwell CD, Asiedu SK, Sibley KJ (2011) Role of ethylene and jasmonic acid on rhizome induction and growth in rhubarb (Rheum rhabarbarum L.). *Plant Cell, Tissue and Organ Culture (PCTOC)* 105(2):253–263.

Reinert J (1959) Phototropism and phototaxis. *Annual Review of Plant Physiology* 10(1):441–458.

Roggemans J, Claes M-C (1979) Rapid clonal propagation of rhubarb by *in vitro* culture of shoot-tips. *Scientia Horticulturae* 11(3):241–246.

Sharma RK, Sharma S (2017) Seed longevity, germination and seedling vigour of Rheum australe D. Don: A step towards conservation and cultivation. *Journal of Applied Research on Medicinal and Aromatic Plants* 5:47–52.

Shibata S (1993) Rheum species (Rhubarb): In vitro Culture and the Production of Sennosides. In: *Medicinal and Aromatic Plants I.V.* Berlin/Heidelberg, Germany: Springer:296–313.

Skoog F, Miller C (1957). Chemical regulation of growth and organ formation in plant tissue cultured *in vitro*. In: *Symposia of the Society for Experimental Biology* 11:118–130.

Steward F, Mapes MO, Mears K (1958) Growth and organized development of cultured cells. II. Organization in cultures grown from freely suspended cells. *American Journal of Botany* 45(10):705–708.

Tali BA, Ganie AH, Nawchoo IA, Wani AA, Reshi ZA (2015) Assessment of threat status of selected endemic medicinal plants using IUCN regional guidelines: A case study from Kashmir Himalaya. *Journal for Nature Conservation* 23:80–89.

Walkey D, Matthews K (1979) Rapid clonal propagation of rhubarb (Rheum rhaponticum L.) from meristem-tips in tissue culture. *Plant Science Letters* 14(3):287–290.

Xu W, Chen G, Li Y, Wang L (2004) Studies on tissue culture technique of {\sl Rheum tanguticum}. *Acta Botanica Boreali-Occidentalia Sinica* 24(9):1734–1738.

8 Conclusions and Future Prospects

Plants have formed the basis of human existence since the Paleolithic age (Solecki 1975), and over a period of time plants with therapeutic significance were documented in the form of various traditional medicine systems (Wani et al. 2013; Pandith et al. 2014). With research advancements made in the field, the cheap herbal extracts have proved safer with fewer side effects. Such extracts have found their use as efficient chemotherapeutic/chemoprotective agents for the effective treatment of a vast array of human ailments (Pandith et al. 2014). Polygonaceae are a large and diverse family treated differently in varied floras, reflecting its complex taxonomic history. It is monophyletic in origin and exhibits extensive plasticity in growth forms (Sanchez and Kron 2009). The family is noticeably eurypalynous/multipalynous with an extensive range of pollen stratifications that offer a great taxonomic significance. Among the vast diversity of angiosperm taxa and within the Polygonaceae family, one of the significant plant species is rhubarb. The genus *Rheum* L. is an extensively diversified and radiated genus bearing 60 congeneric extant species with eight of them found in the Indian subcontinent (Santapau and Henry 1973; Srivastava 2014; Khan et al. 2019). The different species of genus *Rheum* have been reported to grow across the continents of Asia and Europe in 18 countries at an elevation of 500 to 5400 m asl (Wan et al. 2011). The genus has originated mainly in the mountainous regions of the QTP and its adjacent areas and undergone rapid radiations, probably due to immense uplifts of the QTP (Wan et al. 2011; Sun et al. 2012) with remarkable phenotypic diversifications clearly visible as a measure of its adaptation to different habitat alterations (Wan et al. 2014). This radiation of genus *Rheum* may be correlated with the extensive habitat changes, which may not only have promoted rapid allopatric speciation, but also production of new species through polyploidization or diploid-hybridization. Identification of the mode of speciation that contributed most to the diversity of this genus will be an interesting study for future research.

Rhubarb has remained one of the most popular crude drugs due to its efficiency and mildness, with few side effects (Clifford 1992). It has an average lifespan of five to eight years and is often used as food in Europe (Rumpunen and Henriksen 1999) besides having herbal utility in different traditional medical systems in Asia, Europe, and in the America as well (Moore 1997; Maclean and Taylor 2000; Gardener and Ghronicle 2014; Clementi and Misiti 2010). In traditional Chinese medicine, it is used as a decoction of rhizome/roots of different species of *Rheum* under the name "*da huang*" as a laxative and antibacterial agent to improve blood circulation, rinse out the body, and to ease fever (Lai et al. 2015). Rhubarb, besides being traditionally used as laxative, is also used to

DOI: 10.1201/9780429340390-8

treat gout, kidney stones, and liver ailments, viz. jaundice and digestion-related issues, etc. (Ashnagar et al. 2007). Indeed, research interest is growing into highly medicinal herbs around the world along with some herbs in the genus *Rheum* like *R. australe*, *R. tanguticum*, *R. palmatum*, and *R. officinale*, etc. The remedying properties of *Rheum* are indeed attributed to its biologically active secondary chemical constituents, predominantly anthraquinones and stilbenoids, as well as the dietary phytoconstituents (flavonoids) known for their putative health benefits. The biosynthesis of both anthraquinones and flavonoids occur through the polyketide pathway in which type III polyketide synthases (PKSs) act as major multifunctional proteins. Among the diversity of type III PKSs, CHS is the most essential and first committed enzyme (Pandith et al. 2016, 2018, 2020). According to Wan et al. (2011), CHS gene duplication may be correlated with the differentiation of species and lineages within genus *Rheum* and may prove instrumental in species differentiation in this genus with an intricate taxonomic history. According to Zhou et al. (2009), the CHS gene duplication may have acquired and sub-functionalized the modified role in response to stress-response genes like the COR15 gene. Such regulations enable these species to occupy cold and high-altitude niches (Wan et al. 2011). Therefore, studies on the selective evolution and functional divergence of duplicated CHS genes in the entire genus *Rheum* and their novel response mechanisms for strong radiation and low temperature stress need to be carried out.

Based on the China National Knowledge Infrastructure network (http://www.cnki.net), more than 8000 papers have been published in the past few decades on various aspects of rhubarb including its pharmacology, phytochemistry, and anatomy, as well as the genetic diversity of various species. Among these papers, more than 5000 have been published from 2000 to 2010 (Wang et al. 2010). Although a lot is being achieved in terms of research on rhubarb, the estimation of genetic diversity and species delimitation across habitats and especially from the northwest Indian Himalayas is still a concern. Though some efforts have been made for taxonomic circumscription of this genus (Zhou et al. 2020; Liu et al. 2013; Zhou et al. 2017—and references therein), detailed studies both on morphology, DNA barcoding, and other recent trends of molecular systematics of the genuine species of rhubarb across habitats and ecological niches are yet to be conducted on a large scale, which may otherwise yield valuable information for the species delimitation of this genus. Yang et al. (2001) examined the pollen morphology of 40 species of the genus and found that medicinally important species can be palynologically distinguished which were otherwise morphologically and anatomically interwoven (Li and Zhang 1983). While projecting *Rheum* as a natural group with common ancestry, they further proposed a tentative evolutionary trend for different pollen types based on their morphological variations within the genus *Rheum*. With these studies as the basis, the genus has further been divided into eight sections for its diverse morphological traits based on morphology, pollen exine structure, and cpDNA trnL-F region (Wang 1984). However, according to Ferguson and Sang (2001), low-copy nuclear genes with a fast mutation rate are better promising candidates for study in such types of genera to identify

Conclusions and Future Prospects

mode of speciation/hybridization and diversity. Moreover, various marker techniques which include the traditional methods of RAPD, AFLP, RFLP, ISSR, and SSR have been used to understand the genetic diversity and population structure of a particular species. But many researchers still found some limitations with these methods, which opened the door for more advanced techniques of sequencing various genomic DNA fragments, of cpt DNA in particular, that somehow aided in better resolution of the genetic diversity between and within populations, besides the speciation events. For instance, Sun et al. (2012) chose various cpt DNA fragments, viz. rbcL, ndhF, matK, and trnL-trnF, etc., to reconstruct the evolutionary relatedness and evaluate the history of divergence of various species of *Rheum*. Although the matK barcode region has provided promising results, its origin from organellar DNA (with maternal inheritance) has paved the way for an alternate region (like ITS2) which is believed to furnish better information in dealing with such cases as it is inherited from both parents (Chase and Fay 2009). Not only species delimitation, the ITS2 region has also been proposed as a reliable tool for the screening of officinal rhubarb that may help in its quality control and therapeutic applications (Zhou et al. 2017). Indeed, the concerns are increasing regarding the valid identification of different substituents/adulterants of varied drug preparations like da huang at molecular level for overall consumer safety and trade of such herbal preparations in different herbal markets.

Although the cultivation of European rhubarb had declined in the mid-20th century (Hintze, 1951), different cultivars with a similar genotype are still available; for example, the cultivar "early red" is known by many different names. To supplement this chaos, disorganized crossings and an absence of proper pedigrees from the preliminary pollinations has made it nearly impossible to determine the real origin of the vegetable known today as rhubarb (Turner 1938). Overall, the cultivar identification in rhubarb has not been resolved to a satisfactory level. Therefore, an alternative strategy (preferably based on potential molecular markers) is needed to identify a cultivar, assess its genetic relatedness, and to identify the obscure origin of the vegetable-type rhubarb which so far seems largely ambiguous.

The physicochemical structure, nucleotide composition, and functions of chromatin are highly variable. The genus *Rheum* is monobasic with all species having $x = 11$ (Fedorov 1969; Darlington and Wylie 1956). Among cytologically known species of *Rheum*, nearly 40% are tetraploid, 48% diploid, and the remaining 12% have both diploid and tetraploid cytotypes (Saggoo and Farooq 2011; Ruirui et al. 2010). However, in *R. australe*, only diploid cytotypes have been reported bearing chromosome numbers $2n = 22$ (Gohil and Rather 1986; Saggoo and Farooq 2011). Non-essential supernumerary/accessory genetic material in the form of "B" chromosomes in wild populations of *R. tanguticum* has also been reported (Yanping et al. 2011). With diversification in different regions, various species of *Rheum* have accumulated sizeable genetic differences, mainly determined by their mating systems like in other plant species (Wang et al. 2012). Such diversification in different regions has led to species divergence and speciation within this genus leading to complexities in its phylogeny. Also, to confirm whether or not the morphological traits allied to plant "body plans" (decumbent/caulescent and glasshouse) have evolved in parallel in the genus *Rheum*, many attempts had been made with some confirming the occurrence of rapid radiation in *Rheum* that might have been elicited by the widespread uplifts of the QTP, thereby emphasizing the latter's importance (Rowe et al. 2011; Murphy et al. 2001; Sun et al. 2012). However, confirmation of variability in genes expressed in different morphologies of the same phenotype, and ecological adaptation to alpine/arid habitats could be an interesting study. Nevertheless, owing to the utility of karyology in cytotaxonomy, further studies need to be undertaken to actually circumscribe the rhubarb.

Plant genomes contain significant information for comprehending their architecture. However, rhubarb has received little attention at molecular level which would have enabled the study of relative genomics, as well as serving as a valued source for research on overall plant biology. Therefore, it is necessary to employ a datamining framework that integrates different "omics" approaches to efficiently understand, investigate, and further modulate the specialized metabolite pathways in homo- and/or heterologous hosts. In fact, the high-throughput next-generation sequencing (NGS) technologies including RNA-seq that are more efficient and less expensive (Lee and Hong 2019; Jannesar et al. 2020) need to be employed to study the evolutionary origins and ecology of plants like *Rheum*. Also, the techniques of genome editing have revolutionized the research on different aspects of basic and applied biology. Following the emergence of CRISPR–Cas9, genome editing became a commonly used technique to characterize gene function and to refine the desired traits. Importantly, the newly established Cas9 variants, innovative RNA-guided nucleases, and base-editing systems, as well as the DNA-free CRISPR–Cas9 delivery methods, now provide great opportunities for plant genome engineering (Yin et al. 2017). Such techniques are required to be efficiently used in herbs like *Rheum* for drug quality assessment and improvement.

Indeed, the presence of pharmacologically active phytochemicals in *Rheum* has been confirmed beyond doubt. The genus needs to be further exploited across species in order to isolate such biologically active constituents responsible for its therapeutic activity. Hence, extensive research is required to chemically characterize other species and exploit the ways by which such phytochemicals act as well as their therapeutic potential to combat various diseases and to validate the traditional knowledge of rhubarb. As the pharmacological usefulness of rhubarb is explained in the dedicated chapters, with convincing evidence from numerous reports involving animal studies and cell line studies, the health benefits of these pharmacologically active compounds in the general public are limited by their low bioavailability and assurance of possible pharmacological effects on human bodies. Pertinently, rhubarb has also been found to be poisonous and known to contain some toxic chemicals as well. Rhubarb leaves may impair hemostasis causing nausea and vomiting. Contamination of rhubarb rhizome with heavy metals when taken may cause acute diarrhea; intestinal cramping may deplete potassium and potentiate the effects of cardiac glycosides. Its use has been prohibited in children under 12 years, patients with specific organ dysfunction, patients with chronic intestinal inflammation such as gastric or duodenal ulcers, patients with intestinal

Conclusions and Future Prospects

obstruction or ileus, and patients with history of renal stones due to its high oxalate content. Because of it being a uterine stimulant, rhubarb root is not recommended during pregnancy and by nursing women (Tang and Chan 2014). Nonetheless, further studies, addressing the core issues, are required to properly understand and evaluate the efficacy, biological mechanisms, safety, and new potential health benefits of rhubarb in humans.Over the past several decades, some of the species from genus *Rheum* in the habitats where they occur naturally have seen dwindling populations due to different kinds of prevailing natural and anthropogenic pressures. Overexploitation for herbal drug preparations has led to a significant reduction of its populations in natural stands. Consequently, many species of high therapeutic repute in nature—*R. australe*, *R. webbianum*, *R. tanguticum*, and *R. palmatum*— have figured prominently among endangered plant species (Pandith et al. 2014; Rokaya et al. 2012; Pandith et al. 2018; Rashid et al. 2014; Wang et al. 2012). Although, the practice of *in vitro* micro-propagation/regeneration techniques has been employed for some species like *R. palmatum* (Ishimaru et al. 1990; Kasparova and Siatka 2001a, 2001b; Cui et al. 2008), *R. ribes* (Sepehr and Ghorbanli 2005), *R. webbianum* (Rashid et al. 2014), *R. rhaponticum* (Kozak and Salata 2011; Zhao et al. 2006; Roggemans and Claes 1979), *R. rhabarbarum* (Lepse 2007), *R. officinale* (Ji-yong 2010), *R. ribes* (Sepehr and Ghorbanli 2005), *R. emodi* (Lal and Ahuja 1989), *R. tanguticum* (Xu et al. 2004), *R. coreanum* (Mun and Mun 2016), *R. moorcroftianum* (Maithani 2015), and *R. australe* and *R. spiciforme* (by us: data unpublished), the technique is yet to be employed for the conservation of the rest of the species, and chiefly for the large-scale production of therapeutic metabolites. Culture production and transformation with *Agrobacterium rhizogenes* [a gram-negative soil bacteria having capability to transfer T-DNA containing root inducing (Ri)-plasmid] has been achieved in many medicinal plants. These *in vitro* transformation cultures have attracted considerable attention because of their biochemical and genetic stability, rapid growth rates, and ability to synthesize secondary products at levels comparable to those found in the parent plants within a short timespan (Hamill et al. 1987; Signs and Flores 1990; Toivonen 1993; Giri and Narasu 2000). In *Fagopyrum esculentum*, for example, *in vitro* hairy root cultures have been used to examine the biosynthesis of rutin and other polyphenols (Tanaka et al. 1996). In the case of rhubarb, however, the technique is yet to be employed for large-scale production of secondary metabolites and may serve as a basis of a new era of rhubarb research.

Loss of endemic and endangered taxa, which have fundamental ecological and economic implications, is a crucial issue which must be addressed for which all of us need to get involved in one or the other way for future generations to come. In order to fill the huge appetite of the herbal industry, commercialization and privatization should be the priority. Likewise, researchers working on these important plant species should prioritize to develop good agro-techniques for cultivation of *Rheum* species and other endangered species, so that it may become very feasible for farmers/growers to cultivate these important plants and add them to their cash crops. A better way is to invest proper and substantial efforts toward community-managed conservation plans by linking them with the livelihoods of regional people. It would include authentic identification, assessment of population status and demographic changes, if any, documentation of indigenous traditional knowledge, and involvement of regional mountain-based communities in multiple conservation measures which, in turn, would help a great deal in evolving effective and sustainable conservation strategies that could be recommended for implementation through public participation, local/regional stakeholders, and active involvement of relevant government and non-governmental agencies. Further, introducing a livelihood concept vis-à-vis conservation among mountain-dwelling communities has the potential to enhance a sense of belonging among them for the regions (and the regional natural resources they utilize) where they have been living for decades. Moreover, governments need to ensure policy frameworks regulating the cultivation, trade, and value addition, and to initiate steps at social and administrative level for encouraging entrepreneurship in the medicinal plants sector. This sector indeed has an immense potential to serve both as a reservoir of germplasm of such high value flora and to lead the employment/revenue generation for the state and general public (regional people).

REFERENCES

Ashnagar A, Naseri NG, Nasab HH (2007) Isolation and identification of anthralin from the roots of rhubarb plant (Rheum palmatum). *E-Journal of Chemistry* 4(4):546–549.

Chase MW, Fay MF (2009) Ecology. Barcoding of plants and fungi. *Science* 325(5941):682–683.

Clementi EM, Misiti F (2010) Potential health benefits of rhubarb. In: *Bioactive Foods in Promoting Health*. Amsterdam, The Netherlands: Elsevier:407–423.

Clifford M (1992) *Rhubarb: The Wondrous Drug*. Princeton, NJ, USA: Princeton University Press.

Cui Y, Liu X, Han J, Wang B, Guo D (2008) Biotransformation of podophyllotoxin by cell suspension culture and root culture of Rheum palmatum. *Zhongguo zhong Yao za zhi= zhongguo zhongYao zazhi= China Journal of Chinese Materia Medica* 33(9):989–991.

Darlington CD, Wylie AP (1956) *Chromosome Atlas of Flowering Plants. Chromosome Atlas of Flowering Plants*. London: George Alien & Unwin Ltd.

Fedorov AA (1969) *Chromosome numbers of flowering plants*. Leningrad: Academy of Sciences of the USSR, VL Komarov Botanical Institute, Nauka.

Ferguson D, Sang T (2001) Speciation through homoploid hybridization between allotetraploids in peonies (Paeonia). *Proceedings of the National Academy of Sciences of the United States of America* 98(7):3915–3919.

Gardener C, Ghronicle G (2014) Gardener's magazine. *Rhubarb: The Wondrous Drug* 191:321.

Giri A, Narasu ML (2000) Transgenic hairy roots: Recent trends and applications. *Biotechnology Advances* 18(1):1–22.

Gohil R, Rather G (1986) Cytogenetic studies of some members of Polygonaceae of Kashmir. III Rheum L. *Cytologia* 51(4):693–700.

Hamill JD, Parr AJ, Rhodes MJ, Robins RJ, Walton NJ (1987) New routes to plant secondary products. *Nature Bio/Technology* 5(8):800–804.

Hintze, S. (1951) Rhubarb. In: *Svensk vaÈ xtfoÈ raÈdling Del II TraÈdgaÊ rdsvaÈ xterna SkogsvaÈ xterna*. Nature and Culture, Stockholm, 389 ± 91.

Conclusions and Future Prospects

Ishimaru K, Satake M, Shimomura K (1990) Production of (+)-Catechin in Root and Cell Suspension Cultures of Rheum palmatum L. *Plant Tissue Culture Letters* 7(3):159–163.

Jannesar M, Seyedi SM, Jazi MM, Niknam V, Ebrahimzadeh H, Botanga C (2020) A genome-wide identification, characterization and functional analysis of salt-related long non-coding RNAs in non-model plant Pistacia vera L. using transcriptome high throughput sequencing. *Scientific Reports* 10(1):1–23.

Ji-yong J (2010) Tissue culture of rhubarb. *Yinshan Academic Journal* (Natural Science Edition) 2.

Kasparova M, Siatka T (2001a) Effect of chitosan on the production of anthracene derivatives in tissue culture of Rheum palmatum L. *Ceska a Slovenska Farmacie: Casopis Ceske Farmaceuticke Spolecnosti a Slovenske Farmaceuticke Spolecnosti* 50(5):249–253.

Kasparova M, Siatka T (2001b) Effect of the biotic elicitor, Candida utilis, on the production of anthracene derivatives in a tissue culture of Rheum palmatum L. *Ceska a Slovenska Farmacie: Casopis Ceske Farmaceuticke Spolecnosti a Slovenske Farmaceuticke Spolecnosti* 50(1):41–45.

Khan MI, Pandith SA, Ramazan S, Shah MA, Malik AH, Reshi ZA (2019) Rheum moorcroftianum (Polygonaceae) in Kashmir Himalaya. *Phytotaxa* 405(5):269–275.

Kozak D, Salata A (2011) Effect of cytokinins on in vitro multiplication of rhubarb (Rheum rhaponticum L.) 'Karpow Lipskiego' shoots and ex vitro acclimatization and growth. *Acta Scientiarum Polonorum: Hortorum Cultus* 10:75–87.

Lai F, Zhang Y, Xie D-p, Mai S-t, Weng Y-n, Du J-d, Wu G-p, Zheng J-x, Han Y (2015) A systematic review of rhubarb (a Traditional Chinese Medicine) used for the treatment of experimental sepsis. *Evidence-Based Complementary and Alternative Medicine* (https://doi.org/10.1155/2015/131283).

Lal N, Ahuja PS (1989) Propagation of Indian Rhubarh (Rheum emodi Wall.) using shoot-tip and leaf explant culture. *Plant Cell Reports* 8(8):493–496.

Lee D-J, Hong CP (2019) Transcriptome atlas by long-read RNA sequencing: Contribution to a reference transcriptome. In: *Transcriptome Analysis*. London, United Kingdom: IntechOpen.

Lepse L (2007) Comparison of in vitro and Traditional Propagation Methods of Rhubarb (Rheum rhabarbarum) according to Morphological Features and Yield. In: *International Symposium on Acclimatization and Establishment of Micropropagated Plants 812* III(812):265–270.

Li J, Zhang J (1983) Investigation on origin and quality of commodities of rhubarb. *Chinese Journal of Pharmaceutical Analysis* 3(6):333–339.

Liu B-B, Opgenoorth L, Miehe G, Zhang D-Y, Wan D-S, Zhao C-M, Jia D-R, Liu J-Q (2013) Molecular bases for parallel evolution of translucent bracts in an alpine "glasshouse" plant Rheum alexandrae (Polygonaceae). *Journal of Systematics and Evolution* 51(2):134–141.

Maclean W, Taylor K (2000) *Clinical Manual of Chinese Herbal Patent Medicines.* London, United Kingdom: Pangolin Press.

Maithani U (2015) In-vitro Propagation Studies of Rheum moorcroftianum Royle: A Threatened Medicinal plant from Garhwal Himalaya. *International Journal of Current Microbiology and Applied Sciences* 4(6):596–599.

Moore M (1997) *Herbal Formulas for the Clinic and Home.* Bisbee, AZ: Southwest School of Botanical Medicine.

Mun S-C, Mun G-S (2016) Development of an efficient callus proliferation system for Rheum coreanum Nakai, a rare medicinal plant growing in Democratic People's Republic of Korea. *Saudi Journal of Biological Sciences* 23(4):488–494.

Murphy WJ, Eizirik E, Johnson WE, Zhang YP, Ryder OA, O'Brien SJ (2001) Molecular phylogenetics and the origins of placental mammals. *Nature* 409(6820):614–618.

Pandith SA, Dar RA, Lattoo SK, Shah MA, Reshi ZA (2018) Rheum australe, an endangered high-value medicinal herb of North Western Himalayas: A review of its botany, ethnomedical uses, phytochemistry and pharmacology. *Phytochemistry Reviews : Proceedings of the Phytochemical Society of Europe* 17(3):573–609.

Pandith SA, Dhar N, Rana S, Bhat WW, Kushwaha M, Gupta AP, Shah MA, Vishwakarma R, Lattoo SK (2016) Functional promiscuity of two divergent paralogs of type III plant polyketide synthases. *Plant Physiology* 171(4):2599–2619.

Pandith SA, Hussain A, Bhat WW, Dhar N, Qazi AK, Rana S, Razdan S, Wani TA, Shah MA, Bedi Y, Hamid A, Lattoo SK (2014) Evaluation of anthraquinones from Himalayan rhubarb (Rheum emodi Wall. ex Meissn.) as antiproliferative agents. *South African Journal of Botany* 95:1–8.

Pandith SA, Ramazan S, Khan MI, Reshi ZA, Shah MA (2020) Chalcone synthases (CHSs): The symbolic type III polyketide synthases. *Planta* 251(1):15.

Rashid S, Kaloo ZA, Singh S, Bashir I (2014) Callus induction and shoot regeneration from rhizome explants of Rheum webbianum Royle-a threatened medicinal plant growing in Kashmir Himalaya. *Journal of Scientific and Innovative Research* 3(5):515–518.

Roggemans J, Claes M-C (1979) Rapid clonal propagation of rhubarb by in vitro culture of shoot-tips. *Scientia Horticulturae* 11(3):241–246.

Rokaya MB, Münzbergová Z, Timsina B, Bhattarai KR (2012) Rheum australe D. Don: A review of its botany, ethnobotany, phytochemistry and pharmacology. *Journal of Ethnopharmacology* 141(3):761–774.

Rowe KC, Aplin KP, Baverstock PR, Moritz C (2011) Recent and rapid speciation with limited morphological disparity in the genus Rattus. *Systematic Biology* 60(2):188–203.

Ruirui L, Wang A, Tian X, Wang D, Liu J (2010) Uniformity of karyotypes in Rheum (Polygonaceae), a species-rich genus in the Qinghai-Tibetan Plateau and adjacent regions. *Caryologia* 63(1):82–90.

Rumpunen K, Henriksen K (1999) Phytochemical and morphological characterization of seventy-one cultivars and selections of culinary rhubarb (Rheum spp.). *The Journal of Horticultural Science and Biotechnology* 74(1):13–18.

Saggoo MIS, Farooq U (2011) Cytology of Rheum, a vulnerable medicinal plant from Kashmir Himalaya. *Chromosome Botany* 6(2):41–44.

Sanchez A, Kron KA (2009) Phylogenetic relationships of Afrobrunnichia Hutch. & Dalziel (Polygonaceae) based on three chloroplast genes and ITS. *Taxon* 58(3):781–792.

Santapau H, Henry AN (1973) *Dictionary of the Flowering Plants in India.* New Delhi, India: Directorate, Council of Scientific & Industrial Research.

Sepehr MF, Ghorbanli Z (2005) Formation of Catechin in Callus Cultures and Micropropagation of Rheum ribes L. *Pakistan Journal of Biological Sciences* 8(10):1346–1350.

Signs MW, Flores HE (1990) The biosynthetic potential of plant roots. *BioEssays* 12(1):7–13.

Solecki RS (1975) Shanidar IV, a Neanderthal flower burial in northern Iraq. *Science* 190(4217):880–881.

Srivastava R (2014) Family Polygonaceae in India. *Indian Journal of Plant Sciences* 3(2):112–150.

Sun Y, Wang A, Wan D, Wang Q, Liu J (2012) Rapid radiation of Rheum (Polygonaceae) and parallel evolution of morphological traits. *Molecular Phylogenetics and Evolution* 63(1):150–158.

Conclusions and Future Prospects 149

Tanaka N, Yoshimatsu K, Shimomura K, ISHIMARU K (1996) Rutin and other polyphenols in Fagopyrum esculentum hairy roots. Natural medicines= 生薬學雜誌 50(4):269–272.

Tang YL, Chan SW (2014) A review of the pharmacological effects of piceatannol on cardiovascular diseases. *Phytotherapy Research* 28(11):1581–1588.

Toivonen L (1993) Utilization of hairy root cultures for production of secondary metabolites. *Biotechnology Progress* 9(1):12–20.

Turner D (1938) The economic rhubarbs: A historical survey of their cultivation in Britain. *Journal of the Royal Horticultural Society* 63:355–360.

Wan D, Sun Y, Zhang X, Bai X, Wang J, Wang A, Milne R (2014) Multiple ITS copies reveal extensive hybridization within Rheum (Polygonaceae), a genus that has undergone rapid radiation. *PLOS ONE* 9(2):e89769.

Wan D, Wang A, Zhang X, Wang Z, Li Z (2011) Gene duplication and adaptive evolution of the CHS-like genes within the genus Rheum (Polygonaceae). *Biochemical Systematics and Ecology* 39(4–6):651–659.

Wang X, Hou X, Zhang Y, Li Y (2010) Distribution pattern of genuine species of rhubarb as traditional Chinese medicine. *Journal of Medicinal Plants Research* 4(18):1865–1876.

Wang X, Yang R, Feng S, Hou X, Zhang Y, Li Y, Ren Y (2012) Genetic variation in Rheum palmatum and Rheum tanguticum (Polygonaceae), two medicinally and endemic species in China using ISSR markers. *PLOS ONE* 7(12):e51667.

Wang Z (1984) Flora reipublicae Popularis sinicae. *Tomus* 20(2):9–10.

Wani BA, Ramamoorthy D, Rather MA, Arumugam N, Qazi AK, Majeed R, Hamid A, Ganie SA, Ganai BA, Anand R, Gupta AP (2013) Induction of apoptosis in human pancreatic MiaPaCa-2 cells through the loss of mitochondrial membrane potential ($\Delta\Psi$m) by Gentiana kurroo root extract and LC-ESI-MS analysis of its principal constituents. *Phytomedicine* 20(8–9):723–733.

Xu W, Chen G, Li Y, Wang L (2004) Studies on tissue culture technique of {\sl Rheum tanguticum}. *Acta Botanica Boreali-Occidentalia Sinica* 24(9):1734–1738.

Yang M, Zhang D, Zheng J, Liu J (2001) Pollen morphology and its systematic and ecological significance in Rheum (Polygonaceae) from China. *Nordic Journal of Botany* 21(4):411–418.

Yanping H, Wang L, Li YJC (2011) New Occurrence of B Chromosomes in *Rheum Tanguticum* Maxim. Ex Balf.(Polygonaceae). *Caryologia* 64(3):320–324.

Yin K, Gao C, Qiu J-L (2017) Progress and prospects in plant genome editing. *Nature Plants* 3(8):1–6.

Zhao Y, Zhou Y, Grout BW (2006) Variation in leaf structures of micropropagated rhubarb (Rheum rhaponticum L.) PC49. *Plant Cell, Tissue and Organ Culture* 85(1):115–121.

Zhou D, Zhou J, Meng L, Wang Q, Xie H, Guan Y, Ma Z, Zhong Y, Chen F, Liu J (2009) Duplication and adaptive evolution of the COR15 genes within the highly cold-tolerant Draba lineage (Brassicaceae). *Gene* 441(1–2):36–44.

Zhou T, Zhu H, Wang J, Xu Y, Xu F, Wang X (2020) Complete chloroplast genome sequence determination of Rheum species and comparative chloroplast genomics for the members of Rumiceae. *Plant Cell Reports* 39(6):811–824.

Zhou Y, Du X-L, Zheng X, Huang M, Li Y, Wang X-M (2017) ITS2 barcode for identifying the officinal rhubarb source plants from its adulterants. *Biochemical Systematics and Ecology* 70:177–185.

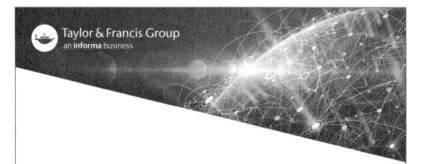